# SEMICONDUCTOR OPTICAL AMPLIFIERS

# Semiconductor Optical Amplifiers

by

**Michael J. Connelly**
*University of Limerick, Ireland*

KLUWER ACADEMIC PUBLISHERS
BOSTON / DORDRECHT / LONDON

A C.I.P. Catalogue record for this book is available from the Library of Congress.

ISBN 0-7923-7657-9

Published by Kluwer Academic Publishers,
P.O. Box 17, 3300 AA Dordrecht, The Netherlands.

Sold and distributed in North, Central and South America
by Kluwer Academic Publishers,
101 Philip Drive, Norwell, MA 02061, U.S.A.

In all other countries, sold and distributed
by Kluwer Academic Publishers,
P.O. Box 322, 3300 AH Dordrecht, The Netherlands.

*Printed on acid-free paper*



Printed in the Netherlands.

For my parents Michael and Margeret
and brother Brendan

# Contents

# Preface

Communications can be broadly defined as the transfer of information from one point to another. In optical fibre communications, this transfer is achieved by using light as the information carrier. There has been an exponential growth in the deployment and capacity of optical fibre communication technologies and networks over the past twenty-five years. This growth has been made possible by the development of new optoelectronic technologies that can be utilised to exploit the enormous potential bandwidth of optical fibre. Today, systems are operational which operate at aggregate bit rates in excess of 100 Gb/s. Such high capacity systems exploit the optical fibre bandwidth by employing wavelength division multiplexing.

Optical technology is the dominant carrier of global information. It is also central to the realisation of future networks that will have the capabilities demanded by society. These capabilities include virtually unlimited bandwidth to carry communication services of almost any kind, and full transparency that allows terminal upgrades in capacity and flexible routing of channels. Many of the advances in optical networks have been made possible by the advent of the optical amplifier.

In general, optical amplifiers can be divided into two classes: optical fibre amplifiers and semiconductor amplifiers. The former has tended to dominate conventional system applications such as in-line amplification used to compensate for fibre losses. However, due to advances in optical semiconductor fabrication techniques and device design, especially over the last five years, the semiconductor optical amplifier (SOA) is showing great promise for use in evolving optical communication networks. It can be utilised as a general gain unit but also has many functional applications including an optical switch, modulator and wavelength converter. These functions, where there is no conversion of optical signals into the electrical domain, are required in transparent optical networks.

It is the intention of this book to provide the reader with a comprehensive introduction to the design and applications of SOAs, particularly with regard to their use in optical communication systems. It is hoped that the book has achieved this aim.

Chapter 1

# INTRODUCTION

In this chapter we begin with the reasons why optical amplification is required in optical communication networks. This is followed by a brief history of semiconductor optical amplifiers (SOAs), a summary of the applications of SOAs and a comparison between SOAs and optical fibre amplifiers (OFAs).

## 1.1 THE NEED FOR OPTICAL AMPLIFICATION

Optical fibre suffers from two principal limiting factors: Attenuation and dispersion. Attenuation leads to signal power loss, which limits transmission distance. Dispersion causes optical pulse broadening and hence intersymbol interference leading to an increase in the system bit error rate (BER). Dispersion essentially limits the fibre bandwidth. The attenuation spectrum of conventional single-mode silica fibre, shown in Fig. 1.1, has a minimum in the 1.55 μm wavelength region. The attenuation is somewhat higher in the 1.3 μm region.

The dispersion spectrum of conventional single-mode silica fibre, shown in Fig. 1.2, has a minimum in the 1.3 μm region. Because the attenuation and material dispersion minima are located in the 1.55 μm and 1.3 μm 'windows', these are the main wavelength regions used in commercial optical fibre communication systems. Systems operating in the 830 nm region are also utilised, mainly for short-haul links at moderate bit rates which do not usually require optical amplification.

Because signal attenuation and dispersion increases as the fibre length increases, at some point in an optical fibre communication link the optical signal will need to be regenerated. 3R (reshaping-retiming-retransmission) regeneration involves detection (photon-electron conversion), electrical amplification, retiming, pulse shaping and retransmission (electron-photon conversion).

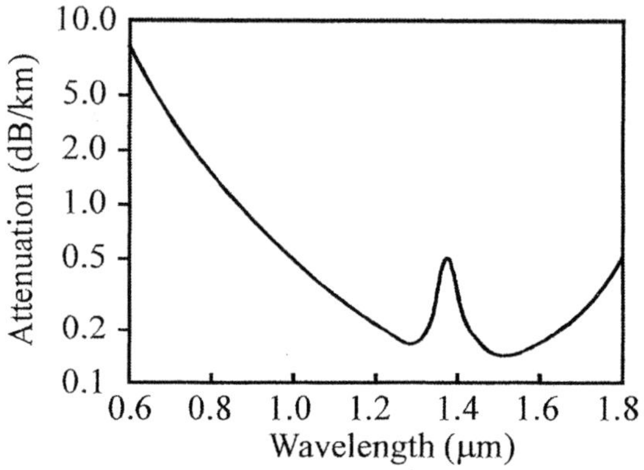


*Figure 1.1* Typical attenuation spectrum of low-loss single-mode silica optical fibre.

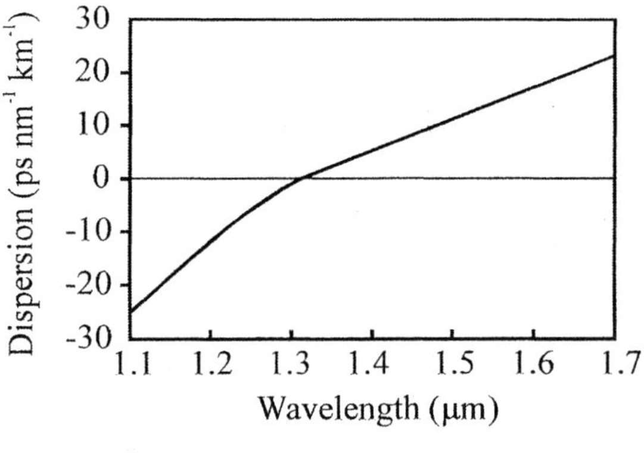


*Figure 1.2* Typical total dispersion spectrum of single-mode silica optical fibre.

This method has a number of disadvantages. Firstly, it involves breaking the optical link and so is not optically transparent. Secondly, the regeneration process is dependent on the signal modulation format and bit rate and so is not electrically transparent. This in turn creates difficulties if the link needs to be upgraded. Ideally link upgrades should only involve changes in or replacement of terminal equipment (transmitter or receiver). Thirdly, as regenerators are complex systems and often situated in remote or difficult to access location, as is the case in undersea transmission links, network reliability is impaired. In systems where fibre loss is the limiting factor, an in-line optical amplifier can be used instead of a regenerator. As the in-line amplifier has only to carry out one function (amplification of the input signal) compared to full regeneration, it is intrinsically a more reliable and less expensive device.

Ideally an in-line optical amplifier should be compatible with single-mode fibre, impart large gain and be optically transparent (i.e. independent of the input optical signal properties).

In addition optical amplifiers can also be useful as power boosters, for example to compensate for splitting losses in optical distribution networks, and as optical preamplifiers to improve receiver sensitivity. Besides these basic system applications optical amplifiers are also useful as generic optical gain blocks for use in larger optical systems. The improvements in optical communication networks realised through the use of optical amplifiers provides new opportunities to exploit the fibre bandwidth.

There are two types of optical amplifier: The SOA and the OFA [1-6]. In recent times the latter has dominated; however SOAs have attracted renewed interest for use as basic amplifiers and also as functional elements in optical communication networks and optical signal processing devices.

## 1.2 BRIEF HISTORY OF SEMICONDUCTOR OPTICAL AMPLIFIERS

The first studies on SOAs were carried out around the time of the invention of the semiconductor laser in the 1960's. These early devices were based on GaAs homojunctions operating at low temperatures. The arrival of double heterostructure devices spurred further investigation into the use of SOAs in optical communication systems. In the 1970's Zeidler and Personick carried out early work on SOAs [7-8]. In the 1980's there were further important advances on SOA device design and modelling. Early studies concentrated on AlGaAs SOAs operating in the 830 nm range [9-10]. In the late 1980's studies on InP/InGaAsP SOAs designed to operate in the 1.3 μm and 1.55 μm regions began to appear [11].

Developments in anti-reflection coating technology enabled the fabrication of true travelling-wave SOAs [12]. Prior to 1989, SOA structures were based on anti-reflection coated semiconductor laser diodes. These devices had an asymmetrical waveguide structure leading to strongly polarisation sensitive gain.

In 1989 SOAs began to be designed as devices in their own right, with the use of more symmetrical waveguide structures giving much reduced polarisation sensitivities [13]. Since then SOA design and development has progressed in tandem with advances in semiconductor materials, device fabrication, antireflection coating technology, packaging and photonic integrated circuits, to the point where reliable cost competitive devices are now available for use in commercial optical communication systems. Developments in SOA technology are ongoing with particular interest in functional applications such as photonic switching and wavelength conversion. The use of SOAs in photonic integrated circuits (PICs) is also attracting much research interest.

## 1.3 SEMICONDUCTOR AND OPTICAL FIBRE AMPLIFIERS: COMPLEMENTARY TECHNOLOGIES

The Erbium Doped Fibre Amplifier (EDFA) was invented in 1985. The EDFA led to a revolution in optical communications as it made possible the replacement of 3R regenerators in links limited by fibre attenuation, leading to the possibility of optically transparent networks and thereby overcoming the ‘electronic bottleneck’. The EDFA has become the optical amplifier of choice in long haul, multichannel digital and analog applications at 1.55 μm. Neodymium Doped Fibre Amplifiers are also available for use in the 1.3 μm region. Fibre amplifiers are especially attractive, as they possess high gain, low insertion loss, low noise figure and negligible nonlinearities, but require an external pump laser.

SOAs have poorer gain and noise and nonlinearities can be severe. However SOA technology is advancing rapidly. SOAs are compatible with monolithic integration (hence low cost potential) and offer a wide range of applications, including optical signal processing that cannot be performed by fibre amplifiers. It is expected that deployment of SOAs in evolving optical communication networks will increase. A comparison between the main features of OFAs and SOAs is given in Table 1.1. Fig. 1.2 illustrates some of diverse range of applications possible with SOAs.

*Table 1.1* Main features of OFAs and SOAs

| Feature | OFA | SOA |
|---|---|---|
| Typical maximum internal gain (dB) | 30 - 50 | 30 |
| Typical insertion loss (dB) | 0.1 - 2 | 6 - 10 |
| Polarisation sensitive? | No | Weak (< 2 dB) |
| Pump source | Optical | Electrical |
| 3 dB gain bandwidth (nm) | 30 | 30 - 50 |
| Nonlinear effects | Negligible | Yes |
| Saturation output power (dBm) | 10 - 15 | 5 - 20 |
| Typical intrinsic noise figure (dB) | 3 - 5 | 7 - 12 dBm |
| Photonic integrated circuit compatible? | No | Yes |
| Functional device possibility? | No | Yes |

It is forecast that the consumption of SOAs will expand rapidly from $48 million in 2000 to $903 million in 2010 [14]. The main near-term use of SOAs will be as basic amplifiers in wavelength division multiplexed (WDM) and other digital optical communication links, and as switching elements in all-optical switches and optical crossconnects. As the above figures show, the future for SOAs is bright, and even more applications of the device will arise as the technology matures and manufacturing costs decrease.

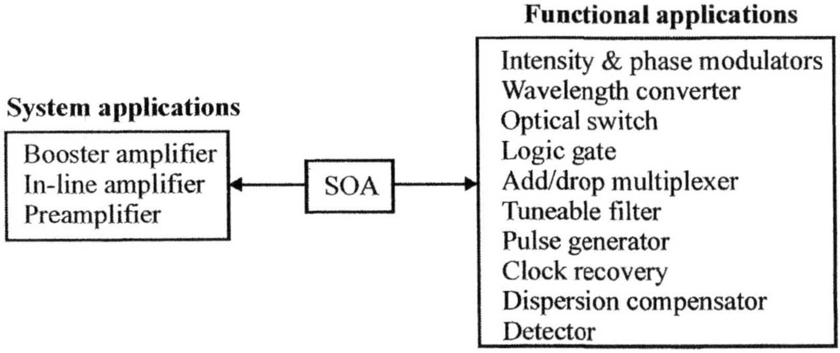


*Figure 1.3.* Typical applications of SOAs.

## REFERENCES


1. S. Shimada and H. Ishio, Eds., *Optical Amplifiers and their Applications*, John Wiley (1992).
2. H. Ghafouri-Shiraz, *Fundamentals of Laser Diode Amplifiers*, John Wiley (1995).
3. Y. Yamamoto, Ed., *Coherence, Amplification and Quantum Effects in Semiconductor Lasers*, John Wiley (1991).
4. E. Desurvire, *Erbium-Doped Fibre Amplifiers: Principles and Applications*, John Wiley, New York (1994).
5. M. J. O'Mahony, Semiconductor optical amplifiers for use in future fibre systems, *IEEE/OSA J. Lightwave Technol.*, **6**, 531-544 (1988).
6. N. A. Olsson, Semiconductor optical amplifiers, *IEEE Proc.*, **80**, 375-382 (1992).
7. G. Zeidler and D. Schicetanz, Use of laser amplifiers in glass fibre communication systems, *Siemens Forch. u. Entwickl. Ber.*, **2**, 227-234 (1973).
8. S. D. Personick, Applications for quantum amplifiers in simple digital optical communication systems, *Bell Syst. Tech. J.*, **52**, 117-133 (1973).
9. Y. Yamamoto, Characteristics of AlGaAs Fabry-Perot cavity type laser amplifiers, *IEEE J. Quantum Electron.*, **16**, 1047-1052 (1980).

10. T. Mukai, Y. Yamamoto and T. Kimura, S/N and error rate performance in AlGaAs semiconductor laser preamplifier and linear repeater systems, *IEEE Trans. Microwave Theory And Tech.*, **30**, 1548-1556 (1982).
11. J. C. Simon, GaInAsP semiconductor laser amplifiers for single-mode fibre communications, *IEEE/OSA J. Lightwave Technol.*, **5**, 1286-1295, 1987.
12. C. E. Zah, C. Caneau, F. K. Shokoohi, S. G. Menocal, F. Favire, L. A. Reith and T. P. Lee, 1.3 μm GaInAsP near-travelling-wave laser amplifiers made by combination of angled facets and antireflection coatings, *Electron. Lett.*, **24**, 1275-1276 (1988).
13. N. A. Olsson, R. F. Kazarinov, W. A. Nordland, C. H. Henry, M. G. Oberg, H. G. White, P. A. Garbinski and A. Savage, Polarisation-independent optical amplifier with buried facets, *Electron. Lett.*, **25**, 1048-1049 (1989).
14. J.D. Montgomery, S. Montgomery and S. Hailu, Semiconductor optical amplifiers expand commercial opportunities, *WDM Solutions, Supplement to Laser Focus World*, 27-30 August 2001.

Chapter 2

# BASIC PRINCIPLES

In this chapter the basic principles of SOAs are explained. The processes of optical gain and additive noise are discussed along with fundamental device parameters including gain ripple, polarisation sensitivity, saturation output power and noise figure.

## 2.1 SOA - BASIC DESCRIPTION

An SOA is an optoelectronic device that under suitable operating conditions can amplify an input light signal. A schematic diagram of a basic SOA is shown in Fig. 2.1. The active region in the device imparts gain to an input signal. An external electric current provides the energy source that enables gain to take place. An embedded waveguide is used to confine the propagating signal wave to the active region. However, the optical confinement is weak so some of the signal will leak into the surrounding lossy cladding regions. The output signal is accompanied by noise. This additive noise is produced by the amplification process itself and so cannot be entirely avoided. The amplifier facets are reflective causing ripples in the gain spectrum.

SOAs can be classified into two main types shown in Fig. 2.2: The Fabry-Perot SOA (FP-SOA) where reflections from the end facets are significant (i.e. the signal undergoes many passes through the amplifier) and the travelling-wave SOA (TW-SOA) where reflections are negligible (i.e. the signal undergoes a single-pass of the amplifier). Anti-reflection coatings can be used to create SOAs with facet reflectivities $< 10^{-5}$. The TW-SOA is not as sensitive as the FP-SOA to fluctuations in bias current, temperature and signal polarisation.

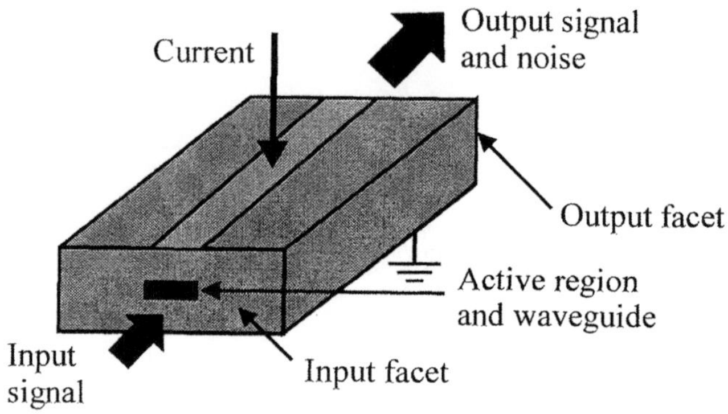


*Figure 2.1.* Schematic diagram of an SOA.

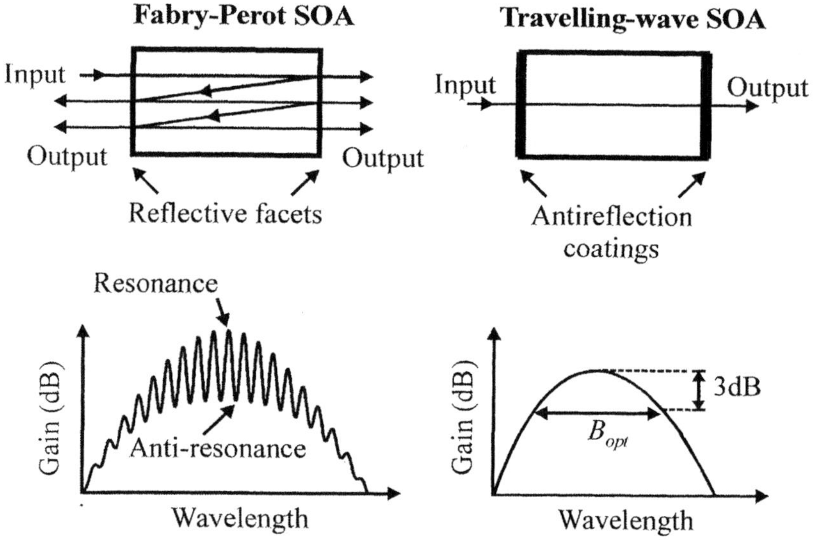


*Figure 2.2.* Basic types of SOA and associated gain spectra. An ideal TW-SOA has a smooth gain spectrum. The FP-SOA gain spectrum exhibits ripples caused by reflections at the end facets. The gain ripples are exaggerated for clarity.

## 2.2 PRINCIPLES OF OPTICAL AMPLIFICATION

In an SOA electrons (more commonly referred to as carriers) are injected from an external current source into the active region. These energised carriers occupy energy states in the conduction band (CB) of the active region material, leaving holes in the valence band (VB). Three radiative mechanisms are possible in the semiconductor. These are shown in Fig 2.3 for a material with an energy band structure consisting of two discrete energy levels.

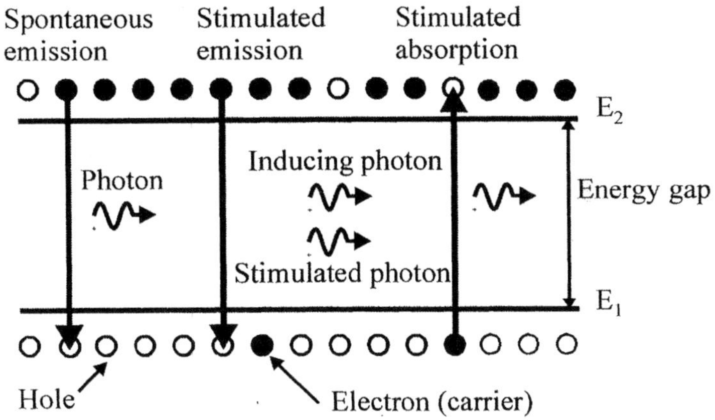


*Figure 2.3* Spontaneous and stimulated processes in a two level system

In stimulated absorption an incident light photon of sufficient energy can stimulate a carrier from the VB to the CB. This is a loss process as the incident photon is extinguished.

If a photon of light of suitable energy is incident on the semiconductor, it can cause stimulated recombination of a CB carrier with a VB hole. The recombining carrier loses its energy in the form of a photon of light. This new stimulated photon will be identical in all respects to the inducing photon (identical phase, frequency and direction, i.e. a coherent interaction). Both the original photon and stimulated photon can give rise to more stimulated transitions. If the injected current is sufficiently high then a population inversion is created when the carrier population in the CB exceeds that in the VB. In this case the likelihood of stimulated emission is greater than stimulated absorption and so semiconductor will exhibit optical gain.

In the spontaneous emission process, there is a non-zero probability per unit time that a CB carrier will spontaneously recombine with a VB hole and thereby emit a photon with random phase and direction. Spontaneously emitted photons have a wide range of frequencies. Spontaneously emitted photons are essentially noise and also take part in reducing the carrier population available for optical gain. Spontaneous emission is a direct consequence of the amplification process and cannot be avoided; hence a noiseless SOA cannot be created. Stimulated processes are proportional to the intensity of the inducing radiation whereas the spontaneous emission process is independent of it.

### 2.2.1 Spontaneous and induced transitions

The gain properties of optical semiconductors are directly related to the processes of spontaneous and stimulated emission. To quantify this relationship we consider a system of energy levels associated with a particular physical system. Let $N_1$ and $N_2$ be the average number of atoms per unit volume of the system characterised by energies $E_1$ and $E_2$ respectively, with $E_2 > E_1$. If a particular atom has energy $E_2$ then there is a finite probability per unit time that it will undergo a transition from $E_2$ to $E_1$, and in the process emit a photon. The spontaneous carrier transition rate per unit time from level 2 to level 1 is given by

$$r_{21}\big|_{spon} = A_{21}N_2 \tag{2.1}$$

where $A_{21}$ is the spontaneous emission parameter of the level 2 to level 1 transition. Along with spontaneous emission it is also possible to have induced transitions. The induced carrier transition rate from level 2 to level 1 (stimulated emission) is given by

$$r_{21}\big|_{stim} = B_{21}\rho(\nu)N_2 \tag{2.2}$$

where $B_{21}$ is the stimulated emission parameter of the level 2 to level 1 transition and $\rho(\nu)$ the incident radiation energy density at frequency $\nu$. The induced photons have energy $h\nu = E_2 - E_1$. The induced transition rate from level 1 to level 2 (stimulated absorption) is given by

$$r_{12} = B_{12}\rho(\nu)N_1 \tag{2.3}$$

where $B_{12}$ is the stimulated emission parameter of the level 2 to level 1 transition. It can be proved, from quantum-mechanical considerations [1,2], that

$$B_{12} = B_{21} \tag{2.4}$$

$$\frac{A_{21}}{B_{21}} = \frac{8\pi n_r^3 h \nu^3}{c^3} \tag{2.5}$$

where $n_r$ is the material refractive index and $c$ the speed of light in a vacuum. Inserting (2.5) into (2.2) gives

$$r_{21}\big|_{stim} = \frac{A_{21} c^3 \rho(\nu) N_2}{8\pi n_r^3 h \nu^3} \tag{2.6}$$

In the case where the inducing radiation is monochromatic at frequency ν, then the induced transition rate from level 2 to level 1 is

$$r_{21}\big|_{stim} = \frac{A_{21} c^3 \rho_\nu l(\nu) N_2}{8\pi n_r^3 h \nu^3} \tag{2.7}$$

where $\rho_\nu$ is the energy density ($J/m^3$) of the electromagnetic field inducing the transition and $l(\nu)$ is the transition lineshape function, normalised such that

$$\int_{-\infty}^{\infty} l(\nu) d\nu = 1 \tag{2.8}$$

$l(\nu)d\nu$ is the probability that a particular spontaneous emission event from level 2 to level 1 will result in a photon with a frequency between $\nu$ and $\nu + d\nu$. The inducing field intensity ($W/m^2$) is

$$I_\nu = \frac{c}{n_r} \rho_\nu \tag{2.9}$$

So (2.7) becomes

$$r_{21}\big|_{stim} = \frac{A_{21} c^2 l(\nu) I_\nu N_2}{8\pi n_r^2 h \nu^3} \tag{2.10}$$

### 2.2.2 Absorption and amplification

By using the expression for the stimulated transition rates developed in Section 2.2.1, it is now possible to derive an equation for the optical gain coefficient for a two level system. We consider the case of a monochromatic plane wave propagating in the *z*-direction through a gain medium with cross-section area *A* and elemental length *dz*. The net power $dP_\nu$ generated by a volume *Adz* of the material is simply the difference in the induced transition rates between the levels multiplied by the transition energy $h\nu$ and the elemental volume i.e.

$$dP_\nu = \left(r_{21}\big|_{stim} - r_{12}\right)h\nu A dz \tag{2.11}$$

This radiation is added coherently to the propagating wave. This process of amplification can then be described by the differential equation

$$\begin{aligned}\frac{dP_\nu(z)}{dz} &= \left(r_{21}\big|_{stim} - r_{12}\right)h\nu A \\ &= g_m(\nu)P_\nu\end{aligned} \tag{2.12}$$

$g_m(\nu)$ is the material gain coefficient given by

$$g_m(\nu) = \frac{A_{21}c^2 l(\nu)(N_2 - N_1)}{8\pi n_r^2 \nu^2} \tag{2.13}$$

(2.13) implies that to achieve positive gain a population inversion ($N_2 > N_1$) must exist between level 2 and level 1. It also shows, by the presence of $A_{21}$, that the process of optical gain is always accompanied by spontaneous emission, i.e. noise. A more detailed description of optical gain in semiconductors is given in Chapter 4.

### 2.2.3 Spontaneous emission noise

As shown above, spontaneous emission is a direct consequence of the amplification process. In this section an expression is derived for the noise power generated by an optical amplifier. We consider the arrangement of Fig. 2.4 [3], which shows an input monochromatic signal of frequency ν travelling through a gain medium having the energy level structure of Fig 2.3. A polariser and optical filter of bandwidth $B_o$ centred about $\nu$ are placed before the detector. The input beam is focussed such that its waist occupies the gain medium. If the beam is assumed to have a *circular* cross-section with waist diameter $D$ then the beam divergence angle is

$$\phi_B = \frac{4\lambda_0}{\pi D n_r} \tag{2.14}$$

where $\lambda_0$ is the free space wavelength. The net change in the signal power due to coherent amplification by an elemental length $dz$ of the gain medium is

$$dP_\nu = g_m(\nu) P_\nu dz \tag{2.15}$$

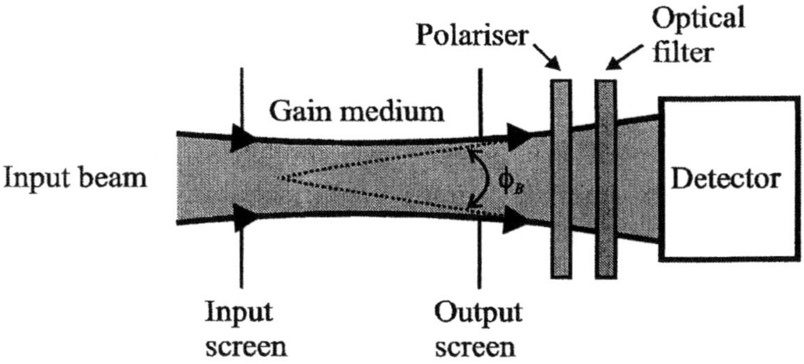


*Figure 2.4.* Optical amplifier consisting of a gain medium contained between two screens. The signal beam is injected so its waist coincides with the input screen (after [3] with the permission of the author).

A volume element, with cross-section area $A$ and length $dz$ at position $z$, of the gain medium spontaneously emits a noise power

$$dP'_N = A_{21} N_2 h\nu A dz \tag{2.16}$$

This noise is emitted isotropically over a 4π solid angle. Each spontaneously emitted photon can exist with equal probability in one of two mutually orthogonal polarisation states. The polariser passes the signal, while reducing the noise by half. Hence the total noise power emitted by the volume element into a solid angle $d\Omega$ and bandwidth $B_o$ is

$$dP_N = \frac{A_{21} N_2 h\nu l(\nu) B_o}{2} \frac{d\Omega A dz}{4\pi} \tag{2.17}$$

The smallest solid angle that can be used without losing signal power [4] is

$$\begin{aligned} d\Omega_{\min} &= \frac{\pi \phi_B^2}{4} \\ &= \frac{\lambda_0^2}{n_r^2 A} \end{aligned} \tag{2.18}$$

This solid angle can be obtained by the use of a suitably narrow output aperture. In this case (2.17) can be rewritten as

$$dP_N = \frac{N_2}{N_2 - N_1} g_m(\nu) h\nu B_o dz \tag{2.19}$$

The total beam power $P$ (signal and noise) can then be described by

$$\frac{dP(z)}{dz} = g_m(\nu) P(z) + n_{sp} g_m(\nu) h\nu B_o \tag{2.20}$$

where the spontaneous emission factor $n_{sp}$ is given by

$$n_{sp} = \frac{N_2}{N_2 - N_1} \tag{2.21}$$

The solution of (2.20), assuming that $g_m$ is independent of $z$, is

$$P(z) = P_{in} e^{g_m z} + n_{sp} h\nu B_o \left(e^{g_m z} - 1\right) \tag{2.22}$$

where $P_{in}$ is the input signal power. If the amplifying medium has length $L$ then the total output power is

$$P_{out} = GP_{in} + n_{sp}(G-1) h\nu B_o \tag{2.23}$$

where $G = e^{g_m L}$ is the single-pass signal gain. The amplifier additive noise power is

$$P_{N,out} = n_{sp}(G-1)h\nu B_o \tag{2.24}$$

(2.24) shows that increasing the level of population inversion can reduce SOA noise. The noise can also be reduced by the use of a narrowband optical filter.

## 2.3 FUNDAMENTAL DEVICE CHARACTERISTICS

The most common application of SOAs is as a basic optical gain block. For such an application, a list of the desired properties is given in Table 2.1. The goal of most SOA research and development is to realise these properties in practical devices.

*Table 2.1* Desirable properties of a practical SOA

| **Property** |
|---|
| High gain and gain bandwidth |
| Negligible facet reflectivities |
| Low polarisation sensitivity |
| High saturation output power |
| Additive noise near the theoretical limit |
| Insensitive to the input signal modulation characteristics |
| Multichannel amplification with no crosstalk |
| No nonlinearities |

### 2.3.1 Small-signal gain and gain bandwidth

In general there are two basic gain definitions for SOAs. The first is the intrinsic gain $G$ of the SOA, which is simply the ratio of the input signal power at the input facet to the signal power at the output facet. The second definition is the fibre-to-fibre gain, which includes the input and output coupling losses. These gains are usually expressed in dB. The gain spectrum of a particular SOA depends on its structure, materials and operational parameters. For most applications high gain and wide gain bandwidth are desired. The small-signal (*small* here meaning that the signal has negligible influence on the SOA gain coefficient) internal gain of a Fabry-Perot SOA at optical frequency $\nu$ is given by [5],

$$G(\nu) = \frac{(1-R_1)(1-R_2)G_s}{\left(1-\sqrt{R_1R_2}G_s\right)^2 + 4\sqrt{R_1R_2}G_S\sin^2[\pi(\nu-\nu_o)/\Delta\nu]} \tag{2.25}$$

where $R_1$ and $R_2$ are the input and output facet reflectivities and $\Delta\nu$ is the cavity longitudinal mode spacing given by

$$\Delta\nu = \frac{c}{2Ln_r} \tag{2.26}$$

$\nu_o$ is the closest cavity resonance to $\nu$. Cavity resonance frequencies occur at integer multiples of $\Delta\nu$. The $sin^2$ factor in (2.25) is equal to zero at resonance frequencies and equal to unity at the anti-resonance frequencies (located midway between successive resonance frequencies). The effective SOA gain coefficient is

$$g = \Gamma g_m - \alpha \tag{2.27}$$

where $\Gamma$ is the optical mode confinement factor (the fraction of the propagating signal field mode confined to the active region) and $\alpha$ the absorption coefficient. $G_s = e^{gL}$ is the single-pass amplifier gain.

An uncoated SOA has facet reflectivities approximately equal to 0.32. The amplifier gain ripple $G_r$ is defined as the ratio between the resonant and non-resonant gains. From (2.25) we get

$$G_r = \left[\frac{1+\sqrt{R_1 R_2}G_s}{1-\sqrt{R_1 R_s}G_s}\right] \tag{2.28}$$

From (2.28) the relationship between the geometric mean facet reflectivity $R_{geo} = \sqrt{R_1 R_2}$ and $G_r$ is

$$R_{geo} = \frac{1}{G_s}\left[\frac{G_r - 1}{G_r + 1}\right] \tag{2.29}$$

Curves of $R_{geo}$ versus $G_s$ are shown in Fig. 2.5 with $G_r$ as parameter. For example, to obtain a gain ripple less than 1 dB at an amplifier single-pass gain of 25 dB requires that $R_{geo} < 3.6 \times 10^{-4}$. Facet reflectivities of this order can be achieved by the application of anti-reflection (AR) coatings to the amplifier facets. The effective facet reflectivities can be reduced further by the use of specialised SOA structures. These techniques are discussed in Chapter 3.

A typical TW-SOA small-signal gain spectrum is shown in Fig. 2.2. The gain bandwidth $B_{opt}$ of the amplifier is defined as the wavelength range over which the signal gain is not less than half its peak value. Wide gain bandwidth SOAs are especially useful in systems where multichannel amplification is required such as in WDM networks. A wide gain bandwidth can be achieved in an SOA with an active region fabricated from quantum-well or multiple quantum-well (MQW) material as discussed in Chapter 4. Typical maximum internal gains achievable in practical devices are in the range of 30 to 35 dB. Typical small-signal gain bandwidths are in the range of 30 to 60 nm.

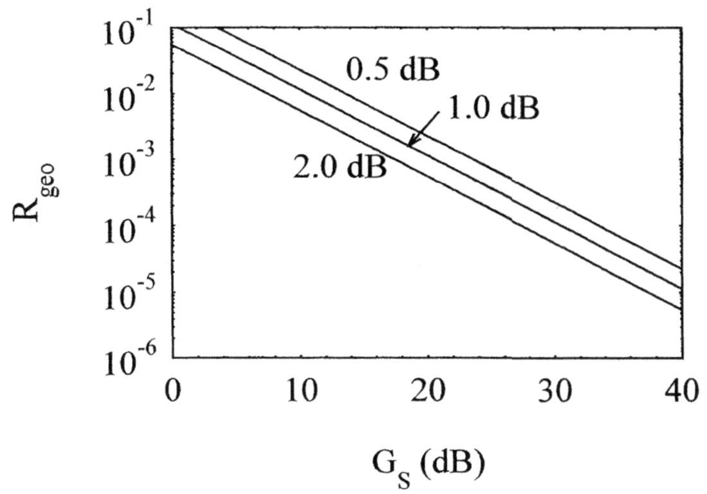


*Figure 2.5.* Geometric mean facet reflectivity, calculated using (2.29), versus single-pass gain with the gain ripple as parameter.

## 2.3.2 Polarisation sensitivity

In general the gain of an SOA depends on the polarisation state of the input signal. This dependency is due to a number of factors including the waveguide structure, the polarisation dependent nature of anti-reflection coatings and the gain material. Cascaded SOAs accentuate this polarisation dependence. The amplifier waveguide is characterised by two mutually orthogonal polarisation modes termed the Transverse Electric (TE) and Transverse Magnetic (TM) modes. The input signal polarisation state usually lies somewhere between these two extremes. The polarisation sensitivity of an SOA is defined as the magnitude of the difference between the TE mode gain $G_{TE}$ and TM mode gain $G_{TM}$, i.e.

$$G_{TE/TM} = |G_{TE} - G_{TM}| \text{ (dB)} \tag{2.30}$$

Techniques for realising SOAs with low polarisation sensitivity (< 1 dB) are discussed in Chapter 3.

### 2.3.3 Signal gain saturation

The gain of an SOA is influenced both by the input signal power and internal noise generated by the amplification process. As the signal power increases the carriers in the active region become depleted leading to a decrease in the amplifier gain. This gain saturation can cause significant signal distortion. It can also limit the gain achievable when SOAs are used as multichannel amplifiers. A typical SOA gain versus output signal power characteristic is shown in Fig. 2.6. A useful parameter for quantifying gain saturation is the saturation output power $P_{o,sat}$, which is defined as the amplifier output signal power at which the amplifier gain is half the small-signal gain. Techniques for realising SOAs with high $P_{o,sat}$ are discussed in Chapter 3. Values in the range of 5 to 20 dBm for $P_{o,sat}$ are typical of practical devices.

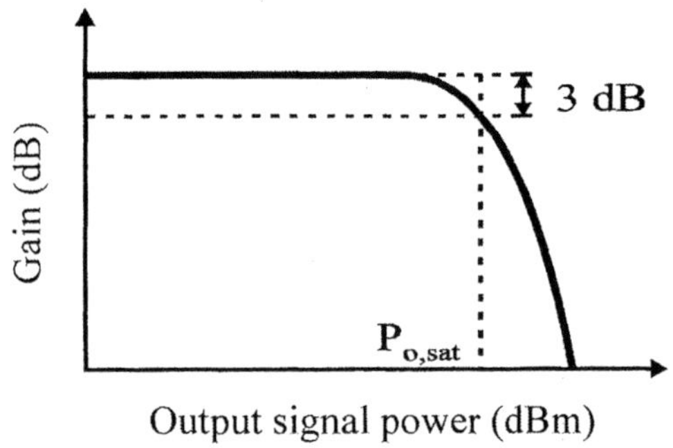


*Figure 2.6* Typical SOA gain versus output signal power characteristic

### 2.3.4 Noise figure

A useful parameter for quantifying optical amplifier noise is the noise figure. $F$, defined as the ratio of the input and output signal to noise ratios, i.e.

$$F = \frac{(S/N)_i}{(S/N)_o} \tag{2.31}$$

The signal to noise ratios in (2.31) are those obtained when the input and output powers of the amplifier are detected by an ideal photodetector. Full expressions for the photocurrent signal to noise ratio in optically amplified systems are derived in Chapter 6.

In the limiting case where the amplifier gain is much larger than unity and the amplifier output is passed through a narrowband optical filter, the noise figure is given by

$$F = 2n_{sp} \tag{2.32}$$

The lowest value possible for $n_{sp}$ is unity, which occurs when there is complete inversion of the atomic medium, i.e. $N_l = 0$, giving $F = 2$ (i.e. 3 dB). Typical intrinsic (i.e. not including coupling losses) noise figures of practical SOAs are in the range of 7 to 12 dB. The noise figure is degraded by the amplifier input coupling loss. Coupling losses are usually of the order of 3 dB, so the noise figure of typical packaged SOAs is between 10 and 15 dB.

### 2.3.5 Dynamic effects

SOAs are normally used to amplify modulated light signals. If the signal power is high then gain saturation will occur. This would not be a serious problem if the amplifier gain dynamics were a slow process. However in SOAs the gain dynamics are determined by the carrier recombination lifetime (average time for a carrier to recombine with a hole in the valence band). This lifetime is typically of a few hundred picoseconds. This means that the amplifier gain will react relatively quickly to changes in the input signal power. This dynamic gain can cause signal distortion, which becomes more severe as the modulated signal bandwidth increases. These effects are further exacerbated in multichannel systems where the dynamic gain leads to interchannel crosstalk. This is in contrast to doped fibre amplifiers, which have recombination lifetimes of the order of milliseconds leading to negligible signal distortion. Dynamic effects are explored further in Chapter 5.

### 2.3.6 Nonlinearities

SOAs also exhibit nonlinear behaviour. In general these nonlinearities can cause problems such as frequency chirping and generation of second or third order intermodulation products. However, nonlinearities can also be of use in using SOAs as functional devices such as wavelength converters. SOA nonlinearities are discussed in more detail in Chapter 7.

## REFERENCES

1. Y. Suematsu and A.R. Adams, *Handbook of Semiconductor lasers and Photonic Integrated Circuits*, London, Chapman and Hall (1994).
2. A. Yariv, *Quantum Electronics*, New York, Wiley (1989).
3. A. Yariv, *Optical Electronics*, New York, HWR International (1985).
4. F.L. Pedrotti and L.S. Pedrotti, *Introduction to Optics*, Prentice-Hall (1993).
5. Y. Yamamoto, Characteristics of AlGaAs Fabry-Perot cavity type laser amplifiers, *IEEE J. Quantum Electron.*, **16**, 1047-1052 (1980).

# Chapter 3

# STRUCTURES

The structural design of an SOA has a large bearing on its performance. SOA structures are chosen so a given device approaches the ideal characteristics listed in Table 2.1 or to accentuate a particular characteristic desirable for a given application (e.g. high saturation output power for booster applications). In this chapter the principles underlying SOA design are reviewed.

## 3.1 SOA BASIC STRUCTURE

In the early days of SOA research, their structures were based on anti-reflection coated double-heterostructure (DH) semiconductor laser diodes as shown in Fig. 3.1. In this type of structure the active region (usually intrinsic i.e. undoped) is sandwiched between n-type and p-type cladding regions. The interfaces between the active region and cladding regions are heterojunctions, as shown in Fig. 3.2. A heterojunction is an interface between two semiconductor materials having different bandgap energies. The cladding regions have higher bandgap energies and lower refractive indices compared to the active region.

Carriers are injected into the device active region from the applied bias current. The injected carriers have to make their way through surrounding layers of semiconductor before they reach the active region. If there were no carrier confinement, the carriers would diffuse throughout the device. As the active region is relatively small, this means that a only a small percentage of the injected carriers would eventually provide gain to a propagating light signal. This leads to inefficient device operation. To overcome this effect, confinement of carriers to the active region is necessary. In the DH structure this is achieved by the bandgap difference between the active region and the cladding regions. This confines carriers to the region between the barriers.

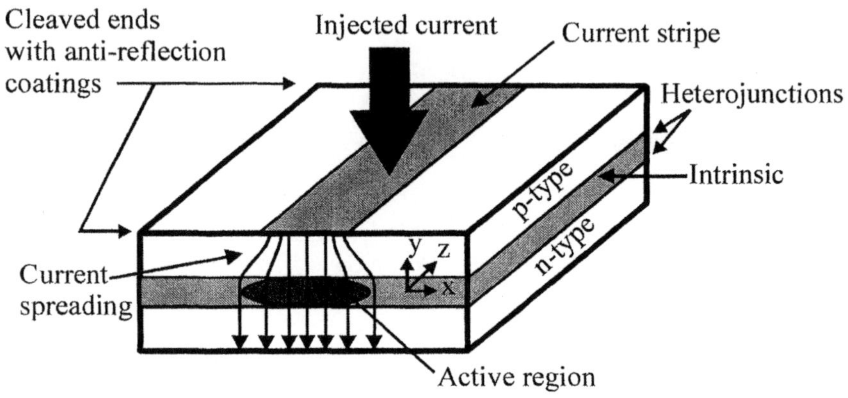


*Figure 3.1* Schematic cross-section of a double-heterostructure SOA

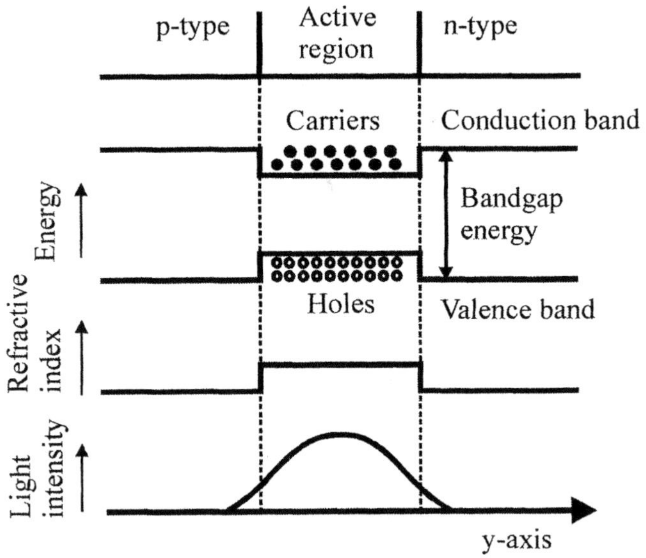


*Figure 3.2.* Carrier and optical confinement in a DH SOA.

However, it is only in the central intrinsic region that the carrier density is high enough to impart gain to a propagating light wave. Carrier confinement has the additional benefit in that the device can be operated at a lower bias current thereby reducing resistive power losses and placing fewer demands on temperature control.

The efficiency of an SOA is improved if the light signal is confined to the active region. In the DH structure the active region has a slightly higher refractive index than the cladding regions and so behaves as a dielectric waveguide with a rectangular cross-section. This helps to confine light travelling through the device to the active region. The amount of waveguiding is quantified by the optical confinement factor $\Gamma$, defined as the fraction of the energy of a particular waveguide mode confined to the active region. Waveguide modes are solutions to Maxwell's equations for the electric and magnetic fields in the waveguide obeying the waveguide boundary conditions. The calculation of dielectric waveguide modes and their associated confinement factors is not a trivial problem; more details can be found in [1-4].

If the waveguide is sufficiently narrow, it will only support a single transverse mode with two possible polarisations, the transverse electric (TE) mode where the electric field is polarised along the heterojunction plane (along the x-axis in Fig. 3.1) and the transverse magnetic (TM) mode where the magnetic field is polarised along the x-axis. The mode is transverse because the associated electric and magnetic fields are both normal to the direction of propagation.

Single transverse mode operation helps to reduce modal gain dependency as the optical confinement factor is mode dependent and also improves the coupling efficiency from the device to optical fibre. The design of an SOA active waveguide to support a single transverse mode is identical to that for laser design. This topic is outside the scope of this book, but is covered in detail in [1-4].

In the above DH SOA there is a well-defined refractive index step in the y-direction between the intrinsic and cladding regions. However, in the x-direction there is no such step. Wave guiding in x-direction is achieved by the injected carriers, which change the refractive index of the intrinsic region. This process is referred to as gain guiding. This refractive index change is less than that in the y-direction. This implies that $\Gamma$ is polarisation dependent. $\Gamma$ increases as the active region thickness is increased. However if the active region is too wide single transverse mode operation will cease. A typical profile of the light field intensity across the device cross-section is shown in Fig. 3.2.

The polarisation dependent optical confinement factors are usually referred to as the TE and TM optical confinement factors, i.e. $\Gamma_{TE}$ and $\Gamma_{TM}$. Methods for reducing polarisation sensitivity include the use of square cross-section active waveguides and strained superlattices (Section 3.3).

## 3.2 SUPPRESSION OF CAVITY RESONANCE

We have seen in Chapter 2 that suppression of SOA facet reflectivities is necessary to achieve travelling-wave operation in an OA. There are a number of methods to reduce the effective facet reflectivities.

### 3.2.1 Antireflection coatings

The power reflectivity for normal incidence at the interface between two dielectrics is given by

$$R = \left( \frac{n_2 - n_1}{n_2 + n_1} \right)^2 \tag{3.1}$$

where $n_1$ and $n_2$ are the dielectric refractive indices. Semiconductor materials have a high refractive index (typically between 3 and 4). Typical cleaved semiconductor-air interface reflectivities are of the order of 32%. While a reflectivity of this magnitude is suitable to achieve laser oscillation in a DH device, it is excessively large for a TW-SOA. The effective facet reflectivities can be greatly reduced by the application of antireflection (AR) coatings. If a plane wave of free-space wavelength $\lambda_0$ is normally incident on a material of refractive index $n_s$ placed in air (refractive index = 1), then the optimal (i.e. for lowest reflectivity) fabrication conditions for an AR coating formed by a single dielectric layer are [5],

$$\begin{aligned} n_f &= \sqrt{n_s} \\ d_f &= \frac{\lambda_0}{4 n_f} \end{aligned} \tag{3.2}$$

where $n_f$ and $d_f$ are the refractive index and thickness of the AR coating. (3.2) only applies to a particular wavelength so a single AR coating is not suitable for SOA operation over a wide bandwidth.

To achieve wideband low facet reflectivities requires the use of multilayer dielectric coatings. The analysis of such coatings is complex and even more difficult when applied to SOAs. This is because the SOA waveguide mode is distributed in the active and cladding regions, which have different refractive indices.

In the following analysis we follow the technique of [6]. If it is assumed that the field distribution is uniform along the direction parallel to the junction between the facet and AR coatings, the waveguide can be analysed using the symmetric two-dimensional model shown in Fig. 3.3.

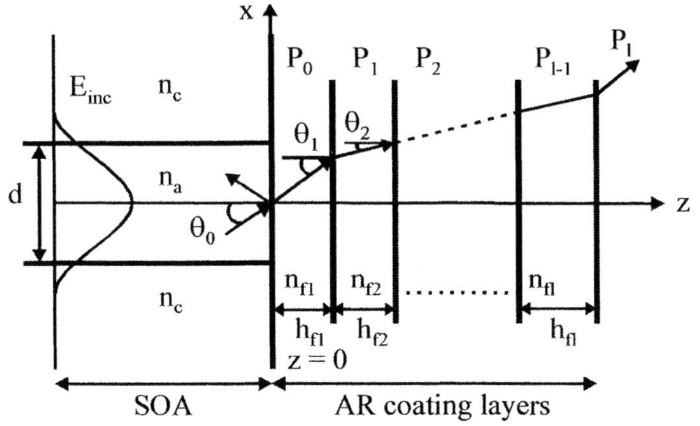


*Figure 3.3* Model of a multilayer AR-coated SOA (z < 0, waveguide in SOA; z > 0, AR coating layers) (after [6] with the permission of the *IEEE*).

In the analysis only TE polarised waves are considered. The active layer has refractive index $n_a$ and thickness $d$. The surrounding cladding regions have refractive index $n_c$ and are assumed to extend to infinity.

The incident field distribution $E_{inc}(x)$ at $z = 0$, the boundary between the active region and the AR coating, is

$$E_{inc}(x) = \begin{cases} A\cos(dh_a/2)\exp[h_c(d/2-|x|)]; & |x| \geq d/2 \\ A\cos(h_a x); \quad |x| \leq d/2 \end{cases} \tag{3.3}$$

where

$$\begin{aligned} h_c^2 &= \beta^2 - k_c^2 \\ h_a^2 &= k_a^2 - \beta^2 \end{aligned} \tag{3.4}$$

which satisfy

$$\tan(dh_a/2) = \frac{h_c}{h_a} \tag{3.5}$$

where the subscripts $a$ and $c$ denote the active and cladding regions respectively. $\beta$ is the propagation constant, $k_a$ ($= 2\pi n_a/\lambda_0$) and $k_c$ ($= 2\pi n_c/\lambda_0$) are wavenumbers and $A$ is an arbitrary constant. The plane wave angular spectrum $F_{inc}(s)$ of the incident field is given by the Fourier transform of (3.3) which gives

$$\begin{aligned} F_{inc}(s) &= F_{inc\_a}(s) + F_{inc\_c}(s) \\ &= A\left\{\frac{\sin[(h_a + k_a s)d/2]}{h_a + k_a s} + \frac{\sin[(h_a - k_a s)d/2]}{h_a - k_a s}\right\} \\ &\quad + \frac{2A}{h_c^2 + k_c^2 s^2}\cos(dh_a/2)\{h_c\cos(dk_c s/2) - k_c s\sin(dk_c s/2)\} \end{aligned} \tag{3.6}$$

where $\theta_o$ is the angle of incidence and $s = sin(\theta_o)$. $F_{inc\_a}$ and $F_{inc\_c}$ are the components of $F_{inc}$ in the active and cladding layers respectively. The Fresnel reflection coefficient of the multilayer AR coating is

$$r(s) = \frac{(m_{11} + m_{12}p_{l+1})p_o - (m_{12} + m_{22}p_{l+1})}{(m_{11} + m_{12}p_{l+1})p_o + (m_{12} + m_{22}p_{l+1})} \tag{3.7}$$

where $l$ is the number of AR coating layers, $p_i = n_{fi}\cos\theta_i \ (i = 0 \cdots l+1)$ and $n_{fj}$ the refractive index of the i-th layer. $n_{f0}$ is the refractive index of the active region or cladding layer. $m_{11}$, $m_{12}$, $m_{21}$, $m_{22}$ are the elements of the 2 x 2 transfer matrix $M$ (= $M_1M_2...M_l$). The elements of the transfer matrix of the i-th layer $M_i$ are

$$\begin{aligned} m_{11}(i) &= m_{22}(i) = \cos\gamma_i \\ m_{12} &= -\frac{j}{p_i}\sin\gamma_j \\ m_{21} &= -jp_i\sin\gamma_i \end{aligned} \tag{3.8}$$

with

$$\gamma_i = \frac{2\pi n_{fi} h_{fi}\cos\theta}{\lambda_0} \tag{3.9}$$

where $j = \sqrt{-1}$ and $h_{fi}$ is the thickness of the i-th layer. The reflected-field angular spectrum $F_{ref}(s)$ is

$$F_{ref}(s) = -\left[r_a(s)F_{inc\_a}(s) + r_c(s)F_{inc\_c}(s)\right] \tag{3.10}$$

where $r_a(s)$ and $r_c(s)$ are the reflection coefficients for the field components in the active and cladding regions respectively. The reflected field $E_{ref}(x)$ at $z = 0$ is equal to the inverse Fourier transform of $F_{ref}(s)$. The reflectivity $R$ for the coated facet is given by the square of the coupling coefficient between $E_{inc}(x)$ and $E_{ref}(x)$, i.e.

$$R = \frac{\left|\int_{-\infty}^{\infty} E_{inc}(x)E_{ref}(x)dx\right|^2}{\left|\int_{-\infty}^{\infty} E_{inc}(x)^2 dx\right|^2} \tag{3.11}$$

This equation can be solved numerically. A similar technique can be applied to obtain the reflectivity of the TM mode. An example of the use of (3.11) for a single layer AR coating is shown in Fig 3.4. With an appropriate combination of film refractive index and thickness, it is possible to achieve an effective facet reflectivity $< 10^{-4}$ using a single layer AR coating.

The AR coating conditions for TE and TM polarisations are not equal. However, the use of multilayer coatings can reduce polarisation dependency and also broaden the low-reflectivity wavelength range [7]. Many dielectric materials such as $SiO_2$, SiN, $SiO_2$-$Si_3N_4$ and PbO-$SiO_2$ can be used as AR coatings. They are applied to an SOA by evaporation or sputtering. The refractive index of the AR coating layer can be controlled by the evaporation or sputtering conditions. Techniques are also available for measuring AR coating reflectivity [8,9].

To achieve low facet reflectivities using AR coatings requires careful control of the refractive index and thickness of the dielectric layers. Altering the SOA structure can reduce the tight tolerances required. Two techniques are commonly used: Angled facets and the window structure, which in conjunction with AR coatings can deliver low reflectivities over a wide bandwidth with minimal polarisation sensitivity.

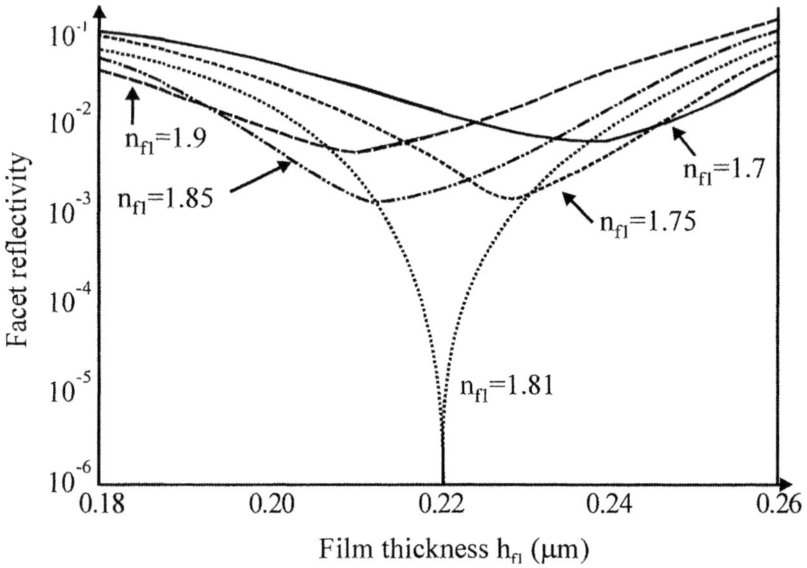


*Figure 3.4.* Single-layer AR coated facet reflectivity versus film thickness with $n_a = 3.524$, $n_c = 3.17$, $d = 0.11$ μm and $\lambda_0 = 1.55$ μm. The parameter is the film refractive index.

### 3.2.2 Angled facet structure

In an angled facet SOA, shown in Fig. 3.5, the active region is slanted away from the facet cleavage plane, thereby reducing the effective facet reflectivity. The V-number of the waveguide is

$$V = \frac{\pi w}{\lambda_0}\sqrt{n_1^2 - n_2^2} \tag{3.12}$$

where $w$ is the waveguide width and $n_1$ and $n_2$ are the effective refractive indices of the active and cladding regions respectively. The TE mode full power width $w_0$ is given by the approximation [10]

$$\frac{w_0}{w} = \frac{1}{\sqrt{2}}\left(9.2063\times10^{-3} + \frac{1.7265}{V} + \frac{0.38399}{V^3} - \frac{9.1691\times10^{-3}}{V^5}\right) \tag{3.13}$$

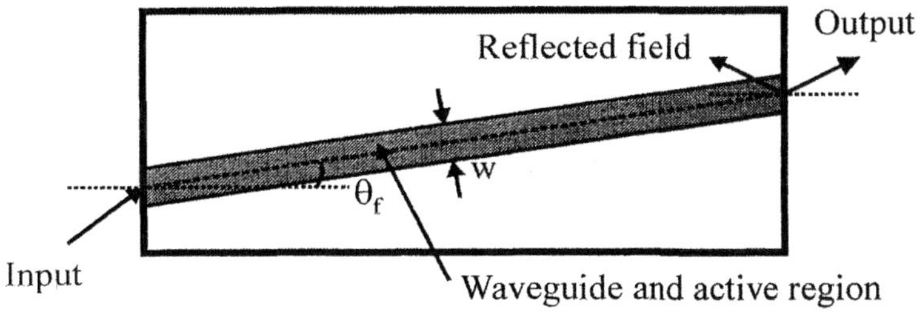


*Figure* 3.5. *Top view of an angled facet SOA.*

If the TE mode is assumed to have a Gaussian distribution, the effective reflectivity of an angled facet is given approximately by

$$R_{ang}(\theta) = R_f(\theta)\exp\left[-\left(\frac{2\pi n_2 w_{full}\theta}{\lambda_0}\right)^2\right] \tag{3.14}$$

where $\theta$ is the angle between the beam propagation direction and the normal to the end facet. The Fresnel reflectivity $R_f$ of a TE plane wave, confined to the waveguide, at the angled facet-air interface is given by

$$R_f(\theta) = \frac{n_1\cos\theta - \sqrt{1-n_1^2\sin^2\theta}}{n_1\cos\theta + \sqrt{1-n_1^2\sin^2\theta}} \tag{3.15}$$

The effective reflectivity of the TM mode is almost identical to that of the TE mode. The effective reflectivity is shown in Fig 3.6 as a function of the facet angle and active region width. The relative reflectivity decreases as the facet angle increases. However the coupling efficiency between an SOA and optical fibre degrades at large facet angles due to the far-field asymmetry. AR coatings also become more polarisation sensitive as the facet angle increases. The relative reflectivity also decreases as the waveguide width increases. However, if the waveguide is too wide higher order transverse modes can appear. This problem can be overcome by broadening the waveguide near the end facets as shown in Fig. 3.7. This also preserves the single transverse mode condition [11]. Optimal facet angles lie in the range $7^0$ to $10^0$. More analyses of angled facets can be found in [12-13].

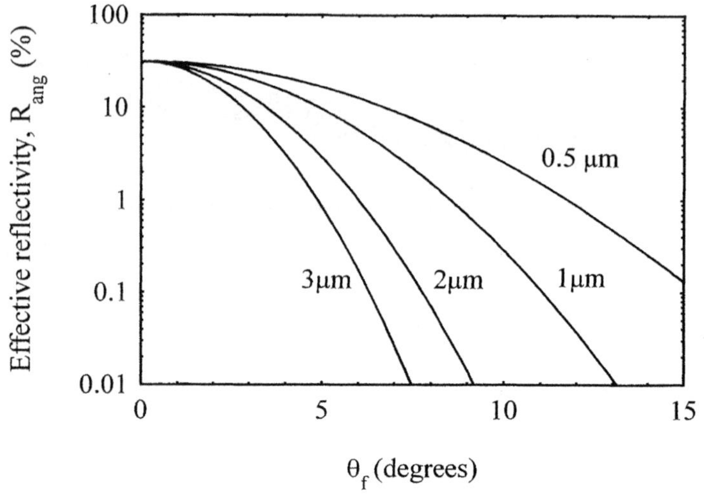


*Figure 3.6* Effective reflectivity of an angled facet as a function of the facet angle with the waveguide width as parameter. The other parameters used are $\lambda_0 = 1.55$ μm, $n_1 = 3.524$ and $n_2 = 3.169$.

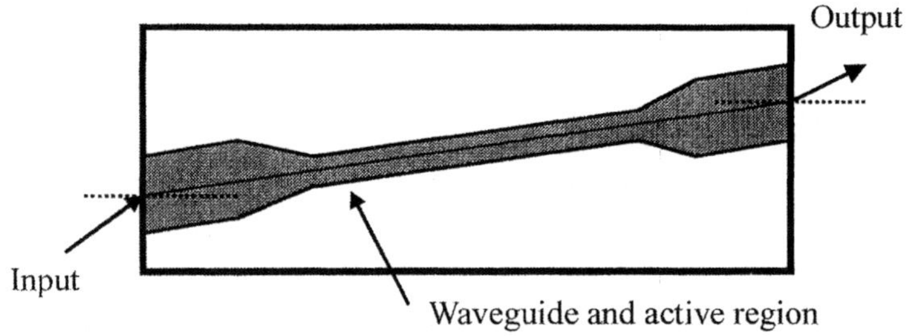


*Figure 3.7.* Top view of an angled-facet flared waveguide SOA (after [11] with permission of the *IEEE*).

### 3.2.3 Window facet structure

As noted above facet reflectivities and optimum AR coating conditions are polarisation dependent. The effective facet reflectivity can be further reduced and made less sensitive to polarisation by the use of window (or buried) facets as shown in Fig 3.8 [14-15]. This structure is simply composed of a transparent region between the active region and end facets. This transparent region has an energy bandgap greater than the signal photon energy. This means that stimulated absorption is not possible although some intrinsic material absorption will be present. The guided field from the waveguide propagates in the window region at some angle due to diffraction and is partially reflected at the end facet. The reflected field continues to broaden in space so only a small fraction is coupled back into the active region. The effective reflectivity decreases with increase in the length $l_w$ of the window region. However, the coupling efficiency from the SOA to an optical fibre is degraded for long $l_w$. Effective facet reflectivities, of the order of 5% are possible with window facets. Used in conjunction with single-layer AR coatings, facet reflectivities $< 10^{-5}$ are possible.

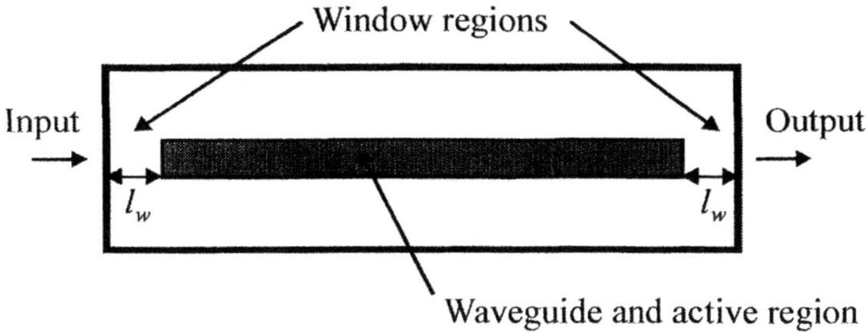


*Figure 3.8.* Top view of an SOA with window facets.

## 3.3 POLARISATION INSENSITIVE STRUCTURES

Polarisation insensitive SOAs are desirable because the polarisation state of the input signal can vary slowly with time. The main cause of polarisation sensitivity is the difference between $\Gamma_{TE}$ and $\Gamma_{TM}$. Polarisation insensitive SOA structural designs aim to reduce or compensate for this difference. In the early stages of SOA development hybrid designs utilising two or more SOA were used to reduce polarisation sensitivity. Those techniques have now been superseded by single chip solutions that mainly focus on improvements in active region design. Three common techniques used are: Square cross-section waveguide, ridge waveguide and strained-layer superlattice material.

### 3.3.1 SOAs with square cross-section active waveguide

Equalisation of $\Gamma_{TE}$ and $\Gamma_{TM}$ can be achieved through the use of a waveguide with square cross-section [16-19] as shown in the buried ridge strip device of

Fig 3.9. In this device the potential barrier of the n-type/p-type InP homojunction is greater than that of the InGaAsP active region/n-type InP heterojunction. This means that there is very little carrier leakage from the active region. This carrier confinement is further improved by the use of highly resistive proton implanted InP regions.

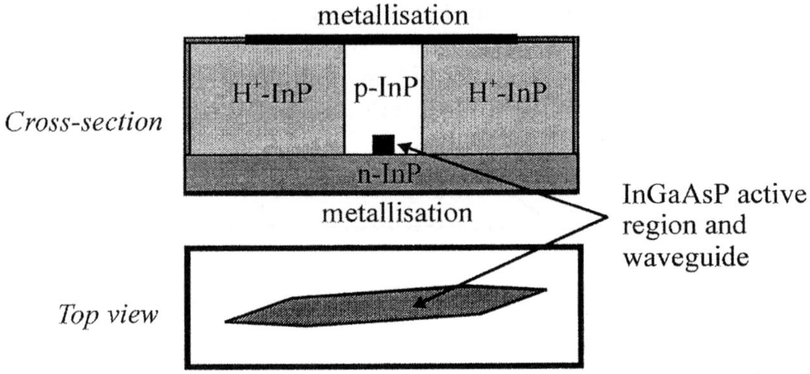


*Figure 3.9* Cross-section and top view of a buried ridge stripe SOA with end tapers and window regions. The central waveguide has a square cross-section with sides of 0.4 μm length (after [16] with the permission of the authors).

However, such structures exhibit large far-field divergence, which leads to poor coupling efficiency from the SOA to optical fibre. Tapering the active waveguide near the amplifier facets as shown in Fig. 3.9 can reduce the far field divergence. [16,18-19]. The guided mode is strongly confined in the central square-cross section waveguide, but experiences less confinement in the tapered sections and so expands. This increases the output mode size and reduces the far-field divergence, thereby increasing the coupling efficiency. The device also includes window regions to reduce the effective facet reflectivity. With this type of device a polarisation sensitivity less than 1 dB was achieved over a wide range of bias currents as shown in Fig. 3.10 [18].

### 3.3.2 Ridge waveguide SOA

The buried ridge waveguide SOA, as shown in Fig. 3.11, has a relatively large active region with a geometry that can be varied to equalise $\Gamma_{TE}$ and $\Gamma_{TM}$ [20]. The relatively large bulk active region and ridge-waveguide structure allow very low modal reflectivities both for AR coated and tilted facets without the requirement for window regions.

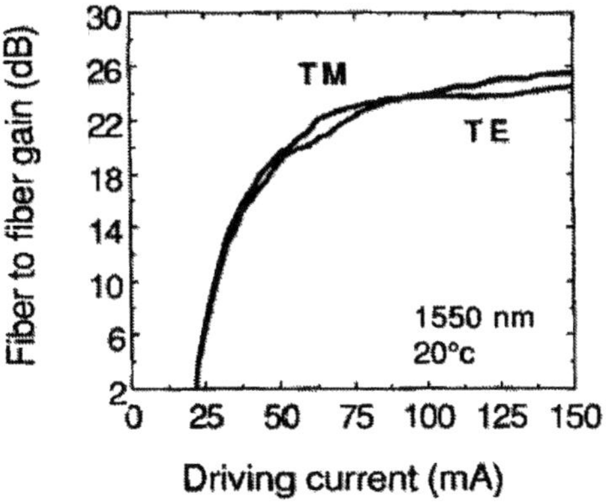


*Figure 3.10.* Fibre-to-fibre gain for TE and TM polarisations versus bias current of a tapered SOA at 1550 nm wavelength (after [18] with permission of the IEEE).

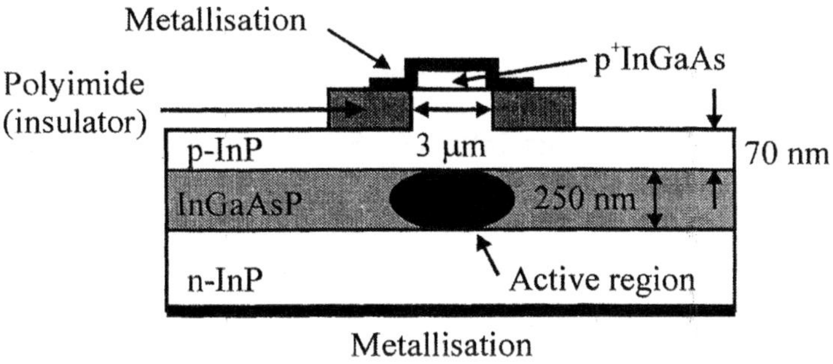


*Figure 3.11.* Bulk ridge-waveguide SOA cross-section (after [20] with permission of the IEEE).

### 3.3.3 Structures based on strained-layer superlattices

If bulk material is used in the active region of an SOA, the only parameter that can be changed to achieve polarisation insensitive operation is the optical confinement factor by using either a square cross-section active waveguide or a ridge-waveguide structure.

Another solution is to keep the usual waveguide geometry (i.e. rectangular cross-section) and to use strained materials in the active region to increase the TM mode gain coefficient relative to the TE mode gain coefficient and thereby compensate for the fact that $\Gamma_{TE} > \Gamma_{TM}$. Reductions in polarisation sensitivity have been reported from device structures with strained tensile barriers [21], tensile strained quantum-wells [22], alternating tensile and compressive strained quantum-wells [23] and strain-balanced superlattices [24]. The particular advantage of the latter structure is that it allows simultaneous control of the polarisation sensitivity and peak gain of the device without limitations placed on the active region thickness. The properties of quantum-wells are covered in more detail in Chapter 4. Using the above techniques low polarisation sensitivity can be achieved over wide bias current and wavelength ranges with the additional advantage of high output saturation power.

## 3.4 HIGH SATURATION OUTPUT POWER STRUCTURES

High saturation output power is a desirable SOA characteristic, particularly for power booster and multichannel applications.

### 3.4.1 Basic model for prediction of amplifier saturation characteristics

To determine the factors that influence SOA gain at high input powers, a simple rate equation model can be used. The amplifier is assumed to have zero facet reflectivities. The material gain coefficient $g_m$ at the signal wavelength is assumed to be a linear function of carrier density $n$,

$$g_m = a_1(n - n_0) \tag{3.16}$$

where $a_l$ is the differential of $g_m$ with respect to $n$ and is assumed here to be a constant. $n_o$ is the transparency carrier density. The carrier density obeys the rate equation

$$\frac{dn}{dt} = \frac{J}{ed} - \frac{n}{\tau} - a_1(n - n_o)\frac{I_S}{h\nu} \tag{3.17}$$

The propagation of the signal intensity $I_s$ through the SOA is described by the travelling-wave equation

$$\frac{dI_s}{dz} = [\Gamma a_1(n - n_o) - \alpha]I_s \tag{3.18}$$

In (3.17) and (3.18), $t$ is time, $z$ is the propagation direction (along the amplifier axis), $J$ is the active region current density, $e$ the electronic charge, $d$ the active region thickness, $\tau$ the spontaneous carrier lifetime, $h$ the Planck constant, $\nu$ the signal optical frequency and $\alpha$ the waveguide loss coefficient. Under steady state conditions the differential in (3.17) is zero. Solving (3.17) in this case gives

$$n = \left(\frac{\tau J}{ed}\right)\left(1 + \frac{I_s}{I_{s,sat}}\right)^{-1} + \frac{n_o I_s}{I_s + I_{s,sat}} \tag{3.19}$$

The saturation intensity $I_{sat}$ and saturation power $P_{sat}$ are given by

$$I_{sat} = \frac{h\nu}{a_1\tau} \qquad P_{sat} = \frac{AI_{sat}}{\Gamma} \tag{3.20}$$

where $A$ is the active region cross-section area. $A/\Gamma$ is the amplifier mode cross-section area. Inserting (3.19) into (3.18) gives

$$\frac{dI_s}{dz} = \left(\frac{\Gamma g_o}{1 + I_s/I_{sat}} - \alpha\right)I_s \tag{3.21}$$

where the unsaturated material gain coefficient $g_o$ is given by

$$g_o = a_1\left(\frac{\tau J}{ed} - n_o\right) \tag{3.22}$$

If for simplicity we assume that $\alpha = 0$, then (3.21) has the solution

$$I_{s,out} \exp\left(\frac{I_{s,out} - I_{s,in}}{I_{s,sat}}\right) = I_{s,in} \exp(\Gamma g_o L) \tag{3.23}$$

where $I_{s,in}$ and $I_{s,out}$ are the input and output signal intensities respectively. The amplifier gain $G$ is the ratio of the output and input signal intensities. From (3.23) we get

$$G = G_o \exp\left[-\frac{(G-1)I_{s,out}}{GI_{sat}}\right] \tag{3.24}$$

where $G_o = \exp(\Gamma g_o L)$ is the unsaturated gain. The amplifier gain, obtained from the numerical solution of (3.24), is shown in Fig. 3.12 as a function of the ratio of $I_{s,out}$ and $I_{sat}$ for unsaturated gains of 10, 20 and 30 dB.

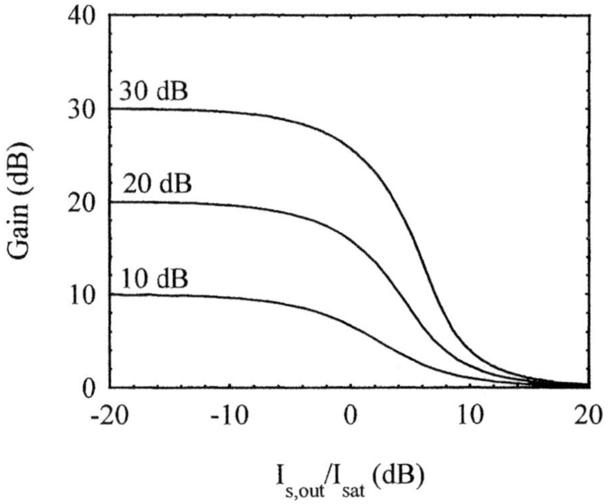


*Figure 3.12.* SOA gain saturation characteristics obtained from the numerical solution of (3.24). The parameter is the unsaturated gain.

From (3.24) the saturation output intensity $\left.I_{s,out}\right|_{3dB}$ (at which the amplifier gain is half the unsaturated gain), is given by

$$\left.I_{s,out}\right|_{3dB} = \frac{\ln(2)G_o I_{s,sat}}{G_o - 2} \tag{3.25}$$

The saturation output power $P_{o,sat}$ of the amplifier is given by

$$P_{o,sat} = \frac{AI_{s,out}\big|_{3dB}}{\Gamma} \tag{3.26}$$

### 3.4.2 Improving saturation output power

(3.26) shows that the saturation output power of an SOA can be improved by increasing $I_{s,sat}$. Inspection of (3.20) shows that this can be achieved by reducing $\tau$ and $a_l$. In practice $\tau$ is inversely proportional to carrier density, so operating at high bias currents leads to an increase in $P_{o,sat}$. However, as the carrier density increases the amplifier gain will also increase making resonance effects more significant. The single-pass gain can be maintained by reducing $\Gamma$ or the amplifier length.. This may not always be necessary as the peak material gain coefficient shifts to shorter wavelengths as the carrier density increases.

The choice of gain material can also influence the saturation behaviour of the amplifier via $a_l$. In bulk materials $a_l$ is relatively sensitive to changes in carrier density. In quantum-well material, conditions can exist where the gain is relatively insensitive to changes in carrier density. This leads to a high $P_{o,sat}$ [25].

It is also possible to increase $P_{o,sat}$ by increasing $A/\Gamma$. An approach based on this concept, shown in Fig. 3.13, is to unfold the amplifier waveguide width towards the output facet [26-28]. This increases the mode field area at the amplifier output.

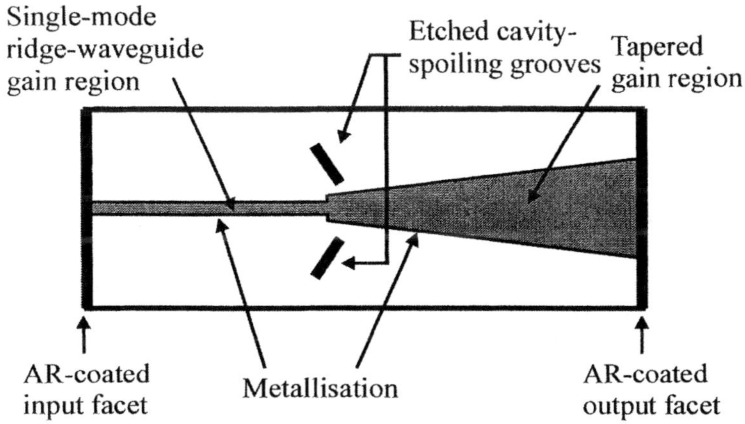


*Figure 3.13* Top view of a 1.3 μm InGaAsP-InP high saturation output power SOA with an input single-mode ridge waveguide region and a tapered gain section. The cavity spoilers reduce the possibility of oscillation (after [27] with permission of the IEEE).

In [27] the device consists of a 1 mm long ridge waveguide input gain section followed by a 2 mm long tapered section. The InGaAsP active region consists of three compressively strained quantum-wells. The device gain was 30 dB at a drive current of 2.8 A with a saturation output power > 720 mW.

Improvement in saturation output power can also be achieved by injecting a continuous wave pump light along with the signal. However the pump signal wavelength has to be chosen so that it lies at the lower edge of the amplifier gain bandwidth. This technique was used in [29] to obtain an improvement of 4.9 dB in the amplifier saturation output power at a signal wavelength of 1546 nm. The pump light had a wavelength of 1480 nm. This technique is not cost effective.

## 3.5 GAIN-CLAMPED SOAs

As Fig. 3.12 shows SOA gain saturation begins to manifest itself at output signal powers well below $P_{o,sat}$ . This can lead to severe crosstalk in multichannel applications. This problem can be much reduced by the use of gain-clamped SOAs (GC-SOAs) [30-33]. In a GC-SOA lasing action is produced, at a wavelength remote from the signal wavelength, by introduction of wavelength specific feedback. Once lasing begins the carrier density is clamped at a fixed value. Changes in the input signal power lead to opposing changes in the lasing mode power. This has the effect of keeping the carrier density fixed (i.e. clamped) making the signal gain relatively insensitive to changes in the input signal power. Common methods of providing this feedback are through the use of a distributed feedback (DFB) structure or distributed Bragg reflectors (DBRs) as shown in Fig. 3.14. A typical gain versus output power characteristic for a GC-SOA is shown in Fig. 3.15 where the signal gain is constant at low powers but falls off rapidly once saturation is reached.

## 3.6 COUPLING LIGHT TO AND FROM SOAs

When SOAs are used in optical transmission systems it is important that there be efficient coupling of light to and from the SOA chip. It is also important that external reflections into the SOA be minimised. A low input coupling loss is also important in order to obtain a low noise figure. Methods to improve coupling efficiency to and from an SOA include the use graded index (GRIN) rod lenses and tapered lens ended optical fibre [34-36]. The use of tapers in the SOA structure can also be used to improve coupling efficiency. The coupling efficiency between an SOA and single-mode optical fibre using the above techniques is typically of the order of 3.5 to 4.5 dB.

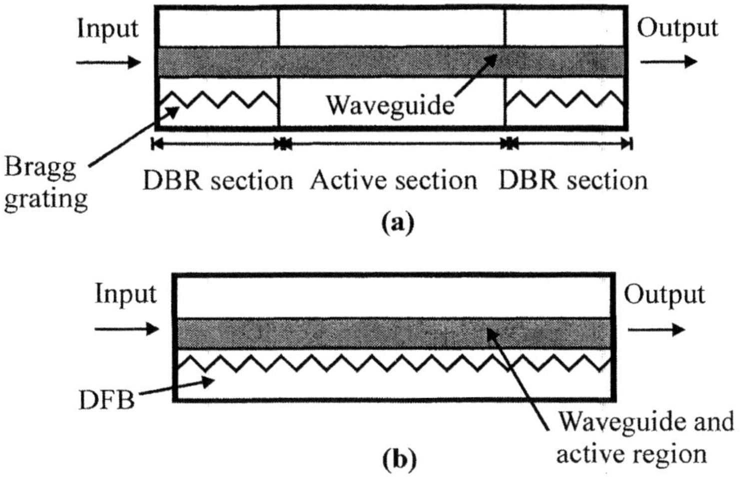


*Figure 3.14.* GC-SOAs with (a) DBR and (b) DFB regions.

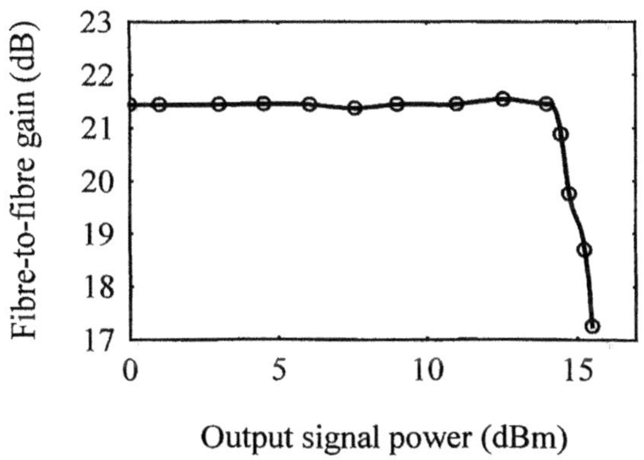


*Figure 3.15.* Typical GC-SOA gain versus output optical power characteristic. The gain curve is linear for output powers below the saturation output power (after [31] with permission of the *IEEE*).

## REFERENCES


1. M.J. Adams, *An Introduction to Optical Waveguides*, John Wiley and Sons (1981).
2. S. L. Chuang, *Physics of Optoelectronic Devices*, John Wiley and Sons, (1995).
3. G. P. Agrawal and N. K. Dutta, *Long-Wavelength Semiconductor Lasers*, Van Nostrand Reinhold, (1986).
4. Y. Suematsu and A.R. Adams, *Handbook of Semiconductor lasers and Photonic Integrated Circuits*, London, Chapman and Hall, (1994).
5. F.L. Pedrotti and L.S. Pedrotti, *Introduction to Optics*, Prentice-Hall, (1993).
6. T. Saitoh, T. Mukai and O. Mikame, Theoretical analysis and fabrication of antireflection coatings on laser-diode facets, *J. Lightwave Technol.*, **3**, 288-293 (1985).
7. C. Vassallo, Polarisation-independent antireflection coatings for semiconductor optical amplifiers, *Electron. Lett.*, **24**, 61-62 (1988).
8. S. A. Merritt, C. Duaga, S. Fox, I.F. Wu and M. Dagenais, Measurement of the facet modal reflectivity spectrum in high quality semiconductor travelling-wave amplifiers, *J. Lightwave Technol.*, **13**, 430-433 (1995).
9. L. Jungkeun and T. Kamiya, Improved characterization of multilayer antireflection coatings for broad-band semiconductor optical amplifiers, *J. Lightwave Technol.*, **18**, 2158 -2166 (2001).
10. D. Marcuse, Reflection loss of laser mode from tilted end mirror, *J. Lightwave Technol.*, **7**, 336-339 (1989).
11. C.E. Zah, R. Bhat, S.G. Menocal, N. Andreadakis, F. Favire, C. Caneau, M.A. Koza and T.P. Lee, 1.5 μm GaInAsP angled-facet flared-waveguide traveling-wave laser amplifiers, *IEEE Photon. Technol. Lett.*, **2**, 46-47 (1990).
12. M. Reed, T.M. Benson, P.C. Kendall and P. Sewell, Antireflection-coated angled facet design, *IEE Proc. Pt. J. Optoelectronics*, **143**, 214 -220 (1996).
13. A. Vukovic, P. Sewell, T.M. Benson and P.C. Kendall, Advances in facet design for buried lasers and amplifiers, *IEEE J. on Sel. Topics in Quantum Electron.*, **6**, 175-184 (2000).
14. I. Cha, M. Kitamura, H. Honmou and I. Mito, 1.5 μm band travelling-wave semiconductor optical amplifier with window facet structure, *Electron. Lett.*, **25**, 1241-1242, (1989).
15. N.A. Olsson, R.F. Kazarinov, W.A. Nordland, C.H. Henry, M.G. Oberg, H.G. White, P.A. Garbinski and A. Savage, Polarisation-independent optical amplifier with buried facets, *Electron. Lett.*, **25**, 1048-1049 (1989).
16. C. Deguet, D. Delprat, G. Crouzel, N.J. Traynor, P. Maigne, T. Pearsal, C. Lerminiaux, N. Andreakis, C. Caneau, F. Favire, R. Bhat, C.E. Zah, Homogeneous Buried Ridge Stripe Semiconductor Optical Amplifier with near Polarisation Independence, *Proc. European Conference on Optical Communications* (1999).
17. B. Mersali, G. Gelly, A. Accard, J.L. Lafragette, P. Doussiere, M. Lambert and B. Fernier, 1.55 μm high-gain polarisation-insensitive semiconductor travelling wave amplifier with low driving current, *Electron. Lett.*, **26**, 124-125 (1990).
18. P. Doussiere, P. Garabedian, C. Graver, D. Bonnevie, T. Fillion, E. Derouin, M. Monnot, J.G. Provost, D. Leclerc, M. Klenk, 1.55 μm polarisation independent semiconductor optical amplifier with 25 dB fiber to fiber gain, *IEEE Photon. Technol. Lett.*, **6**, 170 -172 (1994).
19. D. Tishinin, K. Uppal, I. Kin and P.D. Dapkus, 1.3-μm polarisation insensitive amplifiers with integrated-mode transformers, *IEEE Photon. Technol. Lett.*, **9**, 1337 -1339 (1997).

20. C. Holtmann, P-A. Besse, T. Brenner and H. Melchior, Polarisation independent bulk active region SOAs for 1.3 μm wavelengths, *IEEE Photon. Technol. Lett.*, **8**, 343-345 (1996).
21. M. Magari, M. Okamoto, and Y. Noguchi, 1.55 μm polarisation-insensitive high-gain tensile-strained-barrier MQW optical amplifier, *IEEE Photon. Technol. Lett.*, **3**, 998-1000 (1991).
22. M. Joma, H. Horikawa, C. Q. Xu, K. Yamada, Y. Katoh, and T. Kamijoh, Polarisation insensitive semiconductor laser amplifiers with tensile strained InGaAsP/InGaAsP multiple quantum well structure, *Appl. Phys. Lett.*, **62**, 121-122 (1993).
23. M. A. Newkirk, B. I. Miller, U. Koren, M. G. Young, M. Chien, R. M. Jopson, and C. A. Burrus, 1.5 μm multiquantum-well semiconductor optical amplifier with tensile and compressively strained wells for polarisation independent gain, *IEEE Photon. Technol. Lett.*, **4**, 406-408, (1993).
24. A. Godefroy, A. Le Corre, F. Clerot, S. Salaun, S. Loualiche, J. C. Simon, L. Henry, C. Vaudry, J. C. Keromnes, G. Joulie, and P. Lamouler, 1.55 μm polarisation-insensitive optical amplifier with strain-balanced superlattice active layer, *IEEE Photon. Technol. Lett.*, 7, 473-475 (1995).
25. S. Dubovitsky, W.H. Streifer, A. Mathur and P.D. Dapkus, Gain saturation properties of a semiconductor gain medium with tensile and compressive strain quantum wells, *IEEE J. Quantum Electron.*, **30**, 380-391 (1994).
26. G. Bendelli, K. Komori, S. Arai and Y. Suematsu, A new structure for high-power TW-SLA, *IEEE Photon. Technol. Lett.*, **3**, 42-44 (1991).
27. J.P. Donnelly, J.N. Walpole, G.E. Betts, S.H. Groves, J.D. Woodhouse, F.J. O'Donnell, L.J. Missaggia, R.J. Bailey and A. Napoleone, High-power 1.3 μm InGaAsP-InP amplifiers with tapered gain regions, *IEEE Photon. Technol. Lett.*, **8**, 1450-1542 (1996).
28. H. Ghafouri-Shiraz, P.W. Tan, W.M. Wong, A novel analytical expression of saturation intensity of InGaAsP tapered traveling-wave semiconductor laser amplifier structures, *IEEE Photon. Technol. Lett.*, **10**, 1545 -1547 (1998)
29. M. Yoshino and K. Inoue, Improvement of saturation output power in a semiconductor laser amplifier through pumping light injection, *IEEE Photon. Technol. Lett.*, **8**, 58-59 (1996)
30. B. Bauer, F. Henry and R. Schimpe, Gain stabilization of a semiconductor optical amplifier by distributed feedback, *IEEE Photon. Technol. Lett.*, **6**, 182-185 (1994).
31. L.F. Tiemeijer, G.N. van den Hoven, P.J.A. Thijs, T. van Dongen, J.J.M. Binsma and E.J. Jansen, 1310 nm DBR type MQW gain clamped SOAs with AM-CATV grade linearity, *IEEE Photon. Technol. Lett.*, **8**, 1543-1455 (1996).
32. J. Sum, G. Morthier and R. Baets, Numerical and theoretical study of the crosstalk in gain clamped semiconductor optical amplifiers, *IEEE J. Sel. Topics in Quantum Electron.*, **3**, 1162-1167 (1997).
33. D.T. Schaafsma and E.M. Bradley, Cross-gain modulation and frequency conversion crosstalk effects in 1550-nm gain-clamped semiconductor optical amplifiers, *IEEE Photon. Technol. Lett.*, **11**, 727-729 (1999).
34. T. Toyonaka, S. Tsuji, 22 dB gain semiconductor optical amplifier module using high numerical aperature aspheric lenses, *Electron. Lett.*, **28**, 1302-1303 (1992).
35. T. Toyonaka, S. Tsuji, M. Haneda and Y. Ono, Polarisation-independent semiconductor optical amplifier module using twin GRIN rod lenses, *Electron. Lett.*, **27**, 941 -943 (1991).
36. Z. Wang, B. Mikkelsen, B. Pedersen, K.E. Stubkjaer and D.S. Olesen, Coupling between angled-facet amplifiers and tapered lens-ended fibers, *J. Lightwave Technol.*, **9**, 49 -55 (1991).

# Chapter 4

# MATERIALS

The type of material used in the active region of an SOA will determine its gain spectrum and operational characteristics. In this chapter, the principle properties of common materials used to fabricate SOAs are described. We focus on bulk and quantum-well materials and in particular the InP-InGaAsP and InGaAs-InGaAsP material systems. We use relatively simple models to describe material properties, avoiding the use of complex calculations based on quantum mechanics. A more comprehensive treatment of material properties can be found in [1-8].

SOAs are used mainly in the 1.3 μm and 1.55 μm wavelength regions, and as such are fabricated from the same materials used to fabricate laser sources in these regions. These materials are the III-IV semiconductors, which include InP, InGaAs, InGaAsP, AlGaAs and InAlGaAs.

## 4.1 BULK MATERIAL PROPERTIES

An SOA with an active region whose dimensions are significantly greater than the deBroglie wavelength ($\lambda_B = h/p$ where $p$ is the carrier momentum) of carriers is termed a *bulk* device. In the case where the active region has one or more of its dimensions (usually the thickness) of the order of $\lambda_B$, the SOA is termed a quantum-well (QW) device. It is also possible to have multiple quantum-well (MQW) devices consisting of a number of stacked thin active layers separated by thin barrier (non-active) layers.

### 4.1.1 Bulk material band structure and gain coefficient

The active region of a bulk SOA is fabricated from a direct band-gap material. In such a material the VB maximum and CB minimum energy levels have the same momentum vector. Direct bandgap semiconductors are used because the probability of radiative transitions from the CB to the VB is much greater than is the case for indirect bandgap material. This leads to greater device efficiency, i.e. conversion of injected electrons into photons. The gain model derived here will be used in the SOA models of Chapter 5. A simplified energy band structure of this material type is shown in Fig. 4.1, where there is a single CB and three VBs. The three VBs are the heavy-hole band, light-hole band and a split-off band. The heavy and light-hole bands are degenerate; that is their maxima have the same energy and momentum.

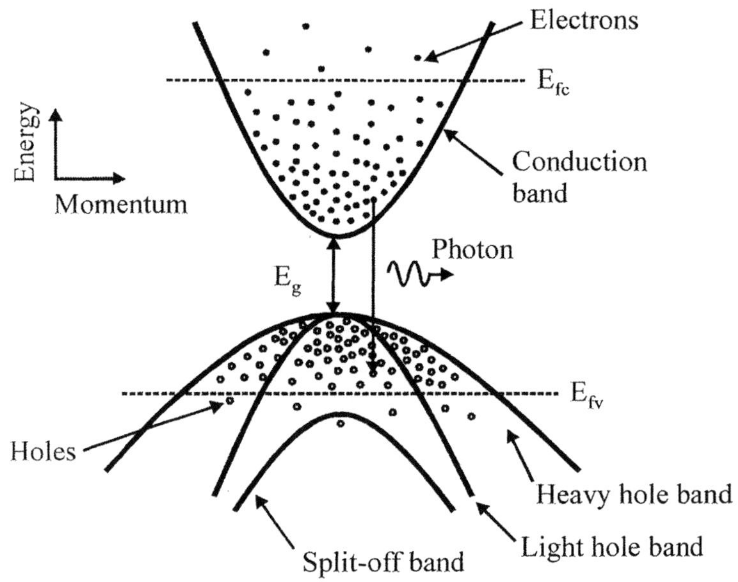


*Figure 4.1.* Simplified energy band structure of a direct bandgap semiconductor. In reality momentum is a vector in three-dimensional space. Only the minority carriers are shown in the conduction and valence bands.

The full description of the band structure is complex; however a good approximation is to assume that the bands have a parabolic shape. In this model the energy of a CB electron or VB hole, measured from the bottom or top of the band respectively is given by

$$E_a = \frac{\hbar^2 k_p^2}{2m_c}$$
$$E_b = \frac{\hbar^2 k_p^2}{2m_v} \tag{4.1}$$

where $k_p$ is the magnitude of the momentum vector, $m_c$ the CB electron effective mass and $m_v$ the VB hole effective mass. Under bias conditions the occupation probability $f_c$ of an electron with energy $E$ in the CB is dictated by Fermi-Dirac statistics given by

$$f_c(E) = \frac{1}{1 + \exp\left(\frac{E - E_{fc}}{kT}\right)} \tag{4.2}$$

where $E_{fc}$ is the quasi-Fermi level of the CB relative to the bottom of the band, $k$ is the Boltzmann constant and $T$ the temperature. Similarly the occupation probability of an electron in the VB with energy $E$, increasing into the band, is given by

$$f_v(E) = \frac{1}{1 + \exp\left(\frac{E - E_{fv}}{kT}\right)} \tag{4.3}$$

where $E_{fv}$ is the quasi-Fermi level of the VB relative to the top of the band. In a parabolic semiconductor the density of allowed electron states is given by

$$\rho_c(E) = \frac{1}{2\pi^2}\left(\frac{2m_c}{\hbar^2}\right)^{3/2}\sqrt{E} \tag{4.4}$$

The CB electron (carrier) density $n$ is equal to the integral over all allowable energies of the product of the density of states and the occupation probability of those states, i.e.

$$n = \frac{1}{2\pi^2}\left(\frac{2m_c}{\hbar^2}\right)^{3/2}\int_0^\infty \frac{\sqrt{E}}{1 + \exp\left[\left(E - E_{fc}\right)/kT\right]}dE \tag{4.5}$$

Similarly the VB hole density $p$ is given by

$$p = \frac{1}{2\pi^2}\left(\frac{2m_v}{\hbar^2}\right)^{3/2}\int_0^\infty \frac{\sqrt{E}}{1+\exp\left(\frac{E-E_{fv}}{kT}\right)}dE \tag{4.6}$$

If $n$ and $p$ are known then $E_{fc}$ and $E_{fv}$ can be obtained by numerical inversion of (4.5) and (4.6). At the high carrier density levels usually present in SOAs $p$ equals $n$. The quasi-Fermi levels can also be estimated using the Nilsson approximation [9]

$$\begin{aligned} E_{fc} &= \left\{\ln\delta + \delta\left[64 + 0.05524\,\delta\left(64+\sqrt{\delta}\right)\right]^{-\frac{1}{4}}\right\}kT \\ E_{fv} &= -\left\{\ln\varepsilon + \varepsilon\left[64 + 0.05524\,\varepsilon\left(64+\sqrt{\varepsilon}\right)\right]^{-\frac{1}{4}}\right\}kT \end{aligned} \tag{4.7}$$

where

$$\delta = \frac{n}{n_c} \quad \text{and} \quad \varepsilon = \frac{p}{n_v} \tag{4.8}$$

$n_c$ and $n_v$ are constants given by

$$\begin{aligned} n_c &= 2\left(\frac{m_e kT}{2\pi\hbar^2}\right)^{3/2} \\ n_v &= 2\left(\frac{m_{dh} kT}{2\pi\hbar^2}\right)^{3/2} \end{aligned} \tag{4.9}$$

and

$$m_{dh} = \left(m_{hh}^{3/2} + m_{lh}^{3/2}\right)^{2/3} \tag{4.10}$$

where $m_{hh}$ and $m_{lh}$ are the VB heavy and light-hole effective masses. For a two-level system we have from (2.13) an expression for the optical gain coefficient at frequency $\nu$,

$$g_m(\nu) = \frac{A_{21}c^2 l(\nu)(N_2 - N_1)}{8\pi n_r^2 \nu^2} \tag{4.11}$$

This expression applies to any *particular* transition. Without lack of generality we can apply it to transitions, having the same momentum vector, between a CB energy level $E_a$ and VB energy level $E_b$ where

$$E_a - E_b = h\nu = E_g + \frac{\hbar^2 k^2}{2m_c} + \frac{\hbar^2 k^2}{2m_v} \tag{4.12}$$

From (4.1) and (4.12) we obtain the relations

$$\begin{aligned} E_a &= (h\nu - E_g)\frac{m_{hh}}{m_c + m_{hh}} \\ E_b &= -(h\nu - E_g)\frac{m_c}{m_c + m_{hh}} \end{aligned} \tag{4.13}$$

In 4.13 it is assumed that heavy-holes dominate over light-holes due to their much greater effective mass. In order to extend (4.11) to a semiconductor material we replace the inversion factor $N_2$ - $N_1$ by the net transition rate from the CB to the VB. The transition rate from energy level $E_a$ in the CB to energy level $E_b$ in the VB, within a frequency interval $d\nu_o$, is

$$dR_{CB \to VB} = \rho_B(\nu_0) f_c(E_a)[1 - f_v(E_b)]\, d\nu_0 \tag{4.14}$$

$\rho_B(\nu_0)$ is the effective joint density of states at transition frequency $\nu_0$, which for a parabolic semiconductor is given by

$$\rho_B(\nu_0) = \frac{1}{\sqrt{\pi}}\left(\frac{2m_c m_{hh}}{\hbar(m_c + m_{hh})}\right)^{3/2}\sqrt{\nu_0 - \frac{E_g}{h}} \tag{4.15}$$

The transition rate in the opposite direction is given by

$$dR_{VB\rightarrow CB} = \rho_B(\nu_0) f_v(E_b)[1 - f_c(E_a)]\, d\nu_0 \tag{4.16}$$

Hence the net transition rate is

$$\begin{aligned} d(N_2 - N_1) &= dR_{CB\rightarrow VB} - dR_{VB\rightarrow CB} \\ &= \rho_B(\nu_0)[f_c(E_a) - f_v(E_b)]\, d\nu_0 \end{aligned} \tag{4.17}$$

Inserting (4.17) into (4.11) and replacing $g_m(\nu_0)$ by $dg_m(\nu_0)$ gives the gain coefficient corresponding to a transition frequency interval $d\nu_0$ as

$$dg_m(\nu) = \frac{A_{21}c^2}{8\pi n_r^2 \nu^2} l(\nu)\rho(\nu_0)[f_c(E_a) - f_v(E_b)]\, d\nu_0 \tag{4.18}$$

To obtain the total gain coefficient we need to integrate (4.18) over the entire range of transition frequencies, so

$$g_m(\nu) = \frac{A_{21}c^2}{8\pi n_r^2 \nu^2} \int_{-\infty}^{\infty} l(\nu)\rho(\nu_0)[f_c(E_a) - f_v(E_b)]\, d\nu_0 \tag{4.19}$$

The transition lineshape function $l(\nu)$ depends on whether the semiconductor material is homogeneously or nonhomogeneously broadened. Homogeneous broadening occurs when each electronic state, is indistinguishable from the others. In this case $l(\nu)$ has a Lorentzian profile

$$l(\nu) = \frac{\Delta\nu}{2\pi\left[(\nu - \nu_0)^2 + (\Delta\nu/2)^2\right]} \tag{4.20}$$

$\Delta\nu$ is the frequency spread of the transition given by

$$\Delta\nu = \frac{1}{\pi\tau_{in}} \tag{4.21}$$

where $\tau_{in}$ is the intraband relaxation time. In the case of nonhomogeneous broadening the transition lineshape has a Gaussian profile. Optical semiconductors are usually homogeneously broadened. In most cases of practical interest the transition lineshape is much narrower than other spectral features of interest in which case $l(\nu)$ can be approximated by a delta function $\delta(\nu - \nu_0)$ within the integral of (4.19). Using this assumption (4.19) becomes

$$g_m(\nu) = \frac{c^2}{4\sqrt{2}\pi^{3/2} n_r^2 \nu^2 \tau_{rad}} \left[ \frac{2m_c m_{hh}}{\hbar(m_c + m_{hh})} \right]^{3/2} \times \left[ f_c(E_a) - f_v(E_b) \right] \sqrt{\nu - \frac{E_g}{h}} \tag{4.22}$$

In (4.22) $A_{21}$ has been replaced by the inverse of $\tau_{rad}$, the radiative carrier recombination lifetime. $\tau_{rad}$ is material dependent and also a function of carrier density. In the above analysis the effect of the split-off band has been neglected.

$g_m$ can be decomposed into a gain coefficient $g'_m(\geq 0)$ and absorption coefficient $g''_m(\geq 0)$,

$$g_m = g'_m - g''_m \tag{4.23}$$

$g'_m$ and $g''_m$ can be obtained by simply replacing $f_c - f_v$ in (4.22) by $f_c[1 - f_v]$ and $f_v[1 - f_c]$ respectively. $g'_m$ is due to stimulated transitions from the CB to the VB. Spontaneous transitions from the CB to the VB have the same spectral shape as $g'_m$. $g''_m$ is due to stimulated transitions from the VB to the CB.

Typical plots of $g_m$, $g'_m$ and $g''_m$ spectra for undoped InGaAsP material are shown in Fig. 4.2. In the calculations the following parameters were used: $E_g = 0.78$ eV, $m_c = 0.046m_0$, $m_{lh} = 0.071m_0$, $m_{hh} = 0.46m_0$, $n_r = 3.7$, $\tau_{rad} = 0.9$ ns and $T = 300$ K. $m_0$ is the electronic rest mass. Plots of $g_m$ spectra with the carrier density as parameter are shown in Fig. 4.3.

### 4.1.2 Analytic approximations to the bulk material gain coefficient

The gain spectrum peak gain and wavelength of InGaAsP material with the above parameters are shown in Fig. 4.4 as a function of carrier density. This shows that the peak gain coefficient is approximately a linear function of the carrier density, i.e.

$$g_{m,peak} \approx a_1(n - n_0) \tag{4.24}$$

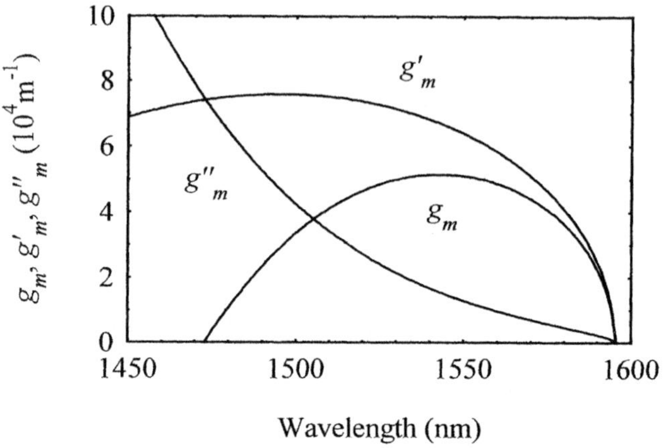


*Figure 4.2* Typical spectral plots of $g_m$, $g'_m$ and $g''_m$ for bulk InGaAsP. The carrier density is 2.0 x $10^{24}$ $m^{-3}$.

where $a_l$ is a constant and $n_0$ is the carrier density at transparency.

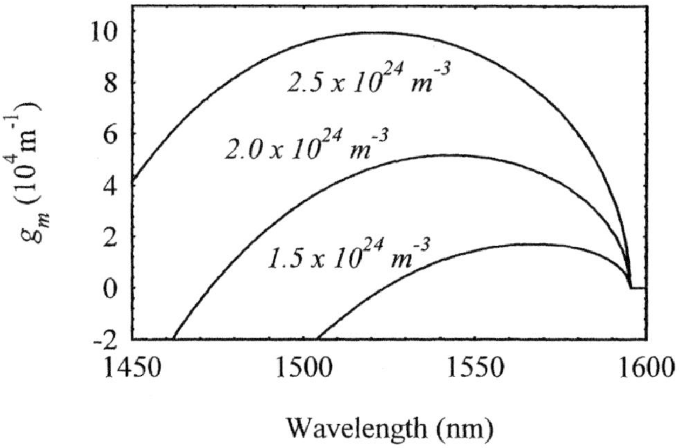


*Figure 4.3.* Typical gain coefficient spectra for bulk InGaAsP. The parameter is carrier density.

For the InGaAsP material gain coefficient shown in Fig. 4.4, $a_l$ = 6.7 x $10^{-20}$ $m^2$ and $n_0$ = 1.2 x $10^{24}$ $m^{-3}$. Fig. 4.4 also shows that the gain peak wavelength $\lambda_{peak}$ shifts to shorter wavelengths as the carrier density increases. The dependency of $\lambda_{peak}$ on the carrier density is also approximately linear, i.e.

$$\lambda_{peak}(n) \approx a_2 n + \lambda_1 \tag{4.25}$$

where $a_2$ and $\lambda_1$ are constants.

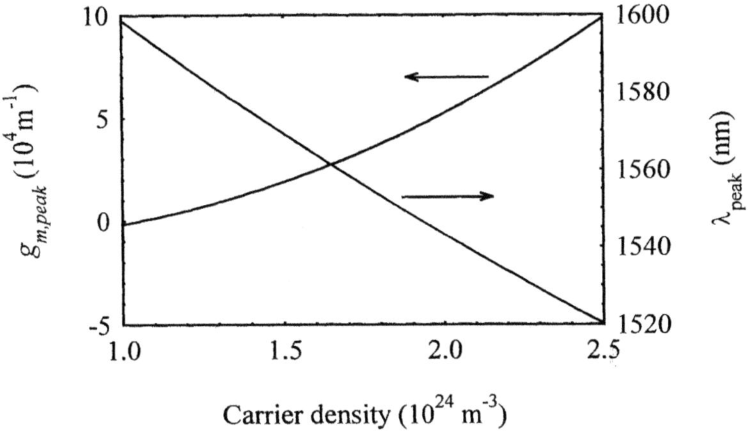


*Figure 4.4* Peak gain coefficient $g_{m,peak}$ and wavelength $\lambda_{peak}$ versus carrier density for bulk InGaAsP.

For the case of Fig. 4.4, $a_2$ = -4.5 x $10^{-23}$ nm/m$^{-3}$ and $\lambda_1$ = 1637 nm. The shape of the gain spectrum is approximately parabolic and can be modelled by

$$g_m(n,\lambda) \approx g_{m,peak}(n) - a_3\left[\lambda - \lambda_{peak}(n)\right]^2 \tag{4.26}$$

$a_3$ is a constant related to the spectral width of the gain spectrum and is given by

$$a_3 \approx \left(\frac{2g_{m,peak}}{\Delta\lambda^2}\right) \tag{4.27}$$

$\Delta\lambda = \lambda_2 - \lambda_1$ where $\lambda_2$ and $\lambda_1$ are the upper and lower wavelengths where $g_m$ is half its peak value. For the gain spectrum of Fig. 4.3 at a carrier density of 2 x $10^{24}$ m$^{-3}$, $\Delta\lambda$ = 95 nm and $a_3$ = 1.14 x $10^{-3}$ m$^{-3}$.

The above approximations are useful in modelling SOAs when quick calculations are required and moderate accuracy is acceptable.

### 4.1.3 Carrier recombination mechanisms in bulk semiconductors

There are two basic types of electron-hole recombination mechanism indirect bandgap optical semiconductors: Radiative and nonradiative.

Radiative recombination occurs when a CB electron recombines with a VB hole leading to the emission of a photon. Radiative recombination can be divided into three processes: Spontaneous emission, stimulated emission and stimulated absorption. The latter two processes have been discussed in relation to optical gain. Spontaneous emission is bimolecular in nature as it involves two states, a CB electron and a VB hole. The spontaneous radiative recombination rate can be modelled [3] as

$$R_{rad} = B_{rad} np \tag{4.28}$$

$B_{rad}$ is the bimolecular radiative recombination coefficient. Typical values of $B_{rad}$ for InGaAsP material are 0.5-2.5 x $10^{-16}$ $m^3s^{-1}$. (4.28) applies to an undoped semiconductor. If the material is doped then

$$R_{rad} = B_{rad} n(p + p_0) \tag{4.29}$$

where $p_0$ is the net ionised carrier concentration, assuming a p-type semiconductor. In SOAs $n$ is usually equal to $p$ so

$$R_{rad} = B_{rad} n(n + p_0) \tag{4.30}$$

The radiative carrier recombination lifetime $\tau_{rad}$, which is used in the gain coefficient calculations of Section 4.2.1, can be modelled as

$$\begin{aligned} \tau_{rad} &\equiv \frac{n}{R_{rad}} \\ &= B_{rad} n \end{aligned} \tag{4.31}$$

Electrons can also recombine with holes nonradiatively. Nonradiative recombination depletes the electron population, which leads to a decrease in the amount of light that can be generated by radiative recombination. Auger recombination is usually the most important nonradiative recombination mechanism. There are many forms of Auger recombination processes [5], an example of which is the CCHC process, illustrated in Fig. 4.4, in which a CB electron and a heavy-hole combine and transfer their excess energy and momentum to a second CB electron, which is excited high into the CB. This process destroys an electron-hole pair.

As the CCHC process involves two CB electrons and one VB hole the Auger recombination rate is [10,11]

$$R_{aug} = C_{aug} n^2 (p + p_0) \tag{4.32}$$

where $C_{aug}$ is the Auger coefficient. Again, in SOAs, $n$ is usually equal to $p$, so

$$R_{aug} = C_{aug} n^2 (n + p_0) \tag{4.33}$$

In general $p_0$ is small compared to $n$ so

$$R_{aug} = C_{aug} n^3 \tag{4.34}$$

This expression holds true for almost all Auger processes. Typical values of $C_{aug}$ for InGaAsP are in the range 0.2 - 2 x $10^{-41}$ $m^6 s^{-1}$ [10].

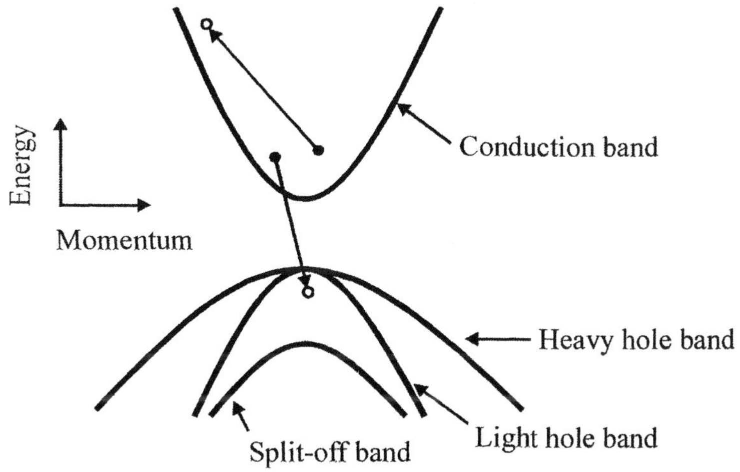


*Figure 4.5.* CCHC Auger recombination process.

Material defects and surface effects can also cause significant nonradiative recombination. Defects (or traps) can occur in the active region of an SOA during the fabrication process and also arise as the device ages. Any carriers close to the trap can recombine nonradiatively. In an SOA the end facets can absorb impurities from the air that can then form nonradiative recombination centres. SOA heterojunctions can also form surfaces for nonradiative recombination. The recombination rate $R_{tr}$ for defect and surface effects is proportional to $n$ and can be modelled be modelled by

$$R_{tr} = A_{tr} n \tag{4.35}$$

where $A_{tr}$ is the trap and surface recombination coefficient. The value of $A_{tr}$ depends on the material quality and the fabrication process. A typical value of $A_{tr}$ for is in the range of $1.0 \times 10^7$ $s^{-1}$. As $R_{tr}$ is linearly proportional to $n$ its effects are only significant when the SOA is operated at low carrier densities.

Another nonradiative recombination mechanism is carrier leakage, where carriers leak across the SOA heterojunctions. This arises due to drift or diffusion of carriers as shown in Fig. 4.6. The leakage recombination rate $R_{leak}$ is given by [10]

$$R_{leak} = \begin{cases} D_{leak} n^{3.5} & \text{diffusion} \\ D_{leak} n^{5.5} & \text{drift} \end{cases} \tag{4.36}$$

depending on whether the diffusion or drift effect dominates. The dominant leakage current is usually due to carrier drift. A typical value of $D_{leak}$ is in the range of $2 \times 10^{48}$ $m^{13.5}s^{-1}$. As the form of (4.36) indicates, carrier leakage effects only become noticeable at high carrier densities.

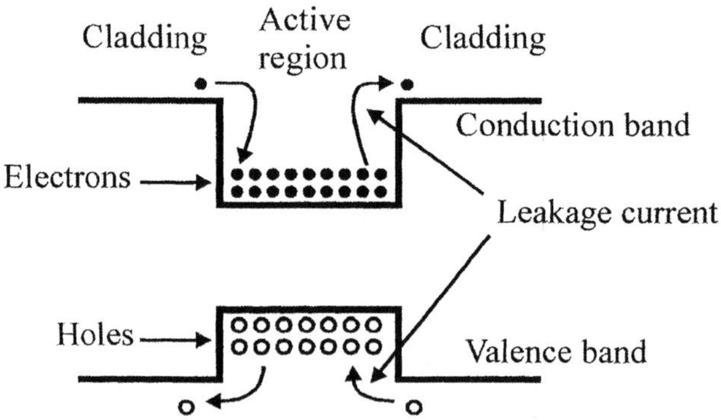


*Figure 4.6.* Heterojunction carrier leakage current

The total nonradiative recombination rate can be written as

$$R_{nrad}(n) = R_{tr}(n) + R_{aug}(n) + R_{leak}(n) \tag{4.37}$$

(4.37) can be expanded as a polynomial in *n*,

$$R_{nrad}(n) = A_{nrad}n + B_{nrad}n^2 + C_{aug}n^3 + D_{leak}n^{5.5} \tag{4.38}$$

where the leakage current is primarily due to carrier drift. The $A_{nrad}$ and $B_{nrad}$ coefficients represent linear and bimolecular nonradiative recombination processes respectively. The total carrier recombination rate *R(n)* is then

$$R(n) = R_{rad}(n) + R_{nrad}(n) \tag{4.39}$$

## 4.2 QUANTUM-WELL MATERIAL PROPERTIES

The active region of a conventional bulk SOA consists of a layer of semiconductor material sandwiched between two cladding layers of higher bandgap energy. If the active layer thickness is less than approximately 20 nm then the occupation states available for confined electrons and holes is no longer continuous but discrete. Such a thin active layer sandwiched between two cladding layers of higher bandgap energy is termed a quantum-well. The active layer is usually referred to as the well layer. An adjacent cladding layer is usually referred to as the barrier layer.

An MQW SOA can be fabricated by stacking well and barrier layers. These very thin layers can be created using fabrication techniques with high controllability such as molecular beam epitaxy (MBE) and organo-metallic vapour phase epitaxy (OMVPE). Compared to bulk SOAs, quantum-well SOAs have wider optical bandwidth, higher saturation output power, and by introducing strain in the material the ability to control polarisation sensitivity. Energy band diagrams of four typical quantum-well structures are shown in Fig. 4.7.

In Fig. 4.7(a) the energy band structure of a conventional single quantum-well (SQW) is shown. This structure has poor carrier and optical confinement. The introduction of a tapered graded index (GRIN) region on either side of the well as shown in Fig. 4.7(b) increases optical confinement. The energy band diagram of a MQW active region is shown in Fig. 4.7(c). In this case because there are more wells the optical gain is increased. In the modified MQW device optical and carrier confinement is improved over the MQW case through the use of an outer cladding region with a higher energy gap than the MQW barrier layers.

### 4.2.1 Unstrained quantum-well band structure and gain coefficient

The mathematics involved in determining the gain coefficient of a semiconductor quantum-well is complex [5,7-8]. In a quantum-well carrier motion in the direction normal to the layer is restricted. This causes quantisation of the well CB and VB into well-defined subbands as shown in Fig. 4.8.

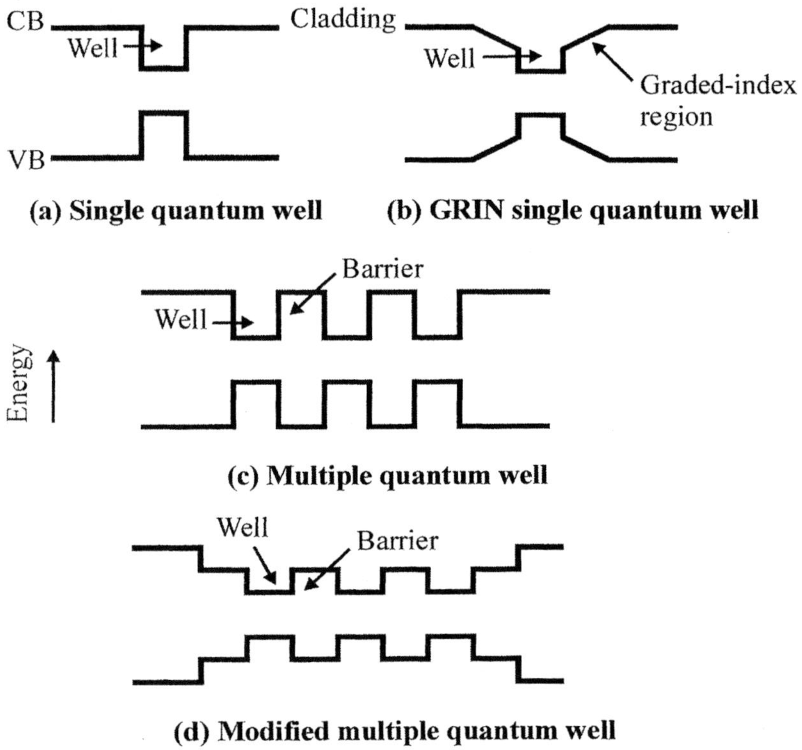


*Figure 4.7.* SQW, GRIN SQW, MQW and modified MQW energy band diagrams.

In Fig. 4.8 $L_z$ is the well thickness, $E_g$ and $E_{gB}$ are the well and barrier energy gaps respectively and $\Delta E_c$ and $\Delta E_v$ are the band edge discontinuities of the CB and VB respectively at the well-barrier interface (heterojunction). $E_{ci}$ is the i-th CB subband. $E_{hhj}$ and $E_{lhj}$ are the j-th quantised heavy and light-hole VB subbands respectively. $E_g$ is the well bandgap energy. As in the bulk material case we assume that the subband energy levels are measured relative to the appropriate band edge and positive into the band.

Using the parabolic band model [12], $E_{ci}$ can be obtained from

$$\frac{m_{cb}}{m_{cw}}\sqrt{\frac{\Delta E_c - E_{ci}}{E_{ci}}} = \begin{Bmatrix} \tan \\ -\cot \end{Bmatrix}\left(\frac{L_z\sqrt{2m_{cw}E_{ci}}}{2\hbar}\right)\begin{Bmatrix} n\,even \\ n\,odd \end{Bmatrix} \tag{4.40}$$

where $m_{cw}$ and $m_{cb}$ are the electron effective masses in the well and barrier respectively. (4.40) generally requires a numerical solution.

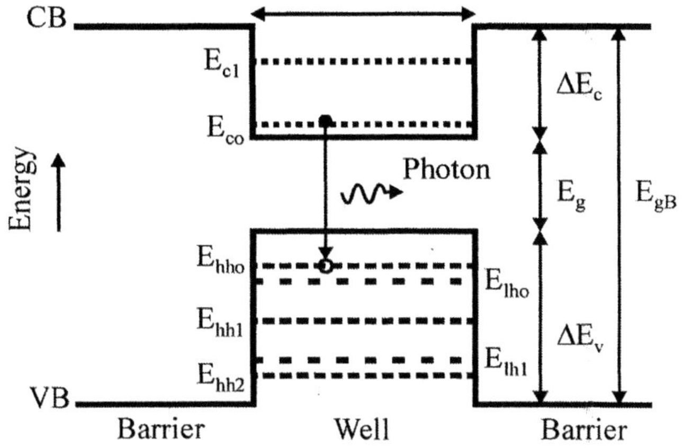


*Figure 4.8.* Energy band model for a single quantum well. The *hh* and *lh* subscripts refer to the heavy- and light-hole subbands respectively.

The maximum number of bound electron states $N_c$ in the well satisfies the condition

$$\frac{L_z\sqrt{2m_{cw}\Delta E_c}}{2\hbar} < \frac{\pi}{2}N_c \tag{4.41}$$

In the case where $E_{cn} << \Delta E_c$, then $E_{cn}$ can be approximated [11] by

$$E_{ci} = \frac{\left[\frac{(i+1)\pi}{2}\frac{a_c}{L_z + \Delta L_c}\right]^2}{\left\{1 + \left[\frac{(i+1)\pi}{2}\right]^2 b_c\left(\frac{\Delta L_c}{L_z + \Delta L_c}\right)^3\right\}} \tag{4.42}$$

where

$$\Delta L_c = \frac{a_c}{\sqrt{b_c \Delta E_c}}$$
$$a_c = \frac{2\hbar}{\sqrt{2m_{cw}}} \tag{4.43}$$
$$b_c = \frac{m_{cw}}{m_{cb}}$$

A similar set of equations can be used to determine $E_{hhi}$ and $E_{lhj}$ by replacing $\Delta E_c$ with $\Delta E_v$, and $m_{cw}$ and $m_{cb}$ substituted by $m_{hhw}$ and $m_{hhb}$ the heavy-hole effective masses in the well and barrier respectively or $m_{lhw}$ and $m_{lhb}$ the light-hole effective masses in the well and barrier respectively.

Under bias conditions the quasi-Fermi levels $E_{fc}$ and $E_{fv}$ in the CB and VB respectively are related to the electron density $n$ and hole density $p$ in the well by

$$n = \frac{m_{cw} kT}{\pi\hbar^2 L_z}\left\{\sum_{i=0}^{N_c-1} \ln\left[1+\exp\left(E_{fc}-E_{ci}\right)/kT\right]\right\}$$
$$p = \frac{kT}{\pi\hbar^2 L_z}\left\{\sum_{j=0}^{N_{hh}-1} m_{hhw} \ln\left[1+\exp\left(E_{fv}-E_{hhj}\right)/kT\right] + \sum_{j=0}^{N_{lh}-1} m_{lhw} \ln\left[1+\exp\left(E_{fv}-E_{lhj}\right)/kT\right]\right\} \tag{4.44}$$

where $N_{hh}$ and $N_{lh}$ are the number of bound heavy-hole and light-hole subbands respectively. As in the bulk case, at the carrier density levels usually present in SOAs, $p$ equals $n$.

The quantum-well gain coefficient can be written as [5],

$$g_m(\nu) = \frac{e^2 E_p}{12\pi\hbar^2 n_r c\varepsilon_0 m_0 L_z \nu}\left\{\sum_{i=0}^{N_c-1}\sum_{j=0}^{N_{hh}-1} m_{hhr} C_{ij} A_{ij} H_{\mathrm{mod}}\left(h\nu - E_{ij}\right) \times\left[f_c(E_{a,ij}) - f_v(E_{b,ij})\right] + \sum_{i=0}^{N_c-1}\sum_{j=0}^{N_{lh}-1} m_{lhr} C_{ij} A_{ij} H_{\mathrm{mod}}\left(h\nu - E_{ij}\right)\left[f_c(E_{a,ij}) - f_v(E_{b,ij})\right]\right\} \tag{4.45}$$

where $E_p$ is the optical matrix parameter and for heavy-holes,

$$\begin{aligned} E_{a,ij} &= \left(h\nu - E_{ij}\right)\frac{m_{hhr}}{m_{cw}} + E_{ci} \\ E_{b,ij} &= -\left(h\nu - E_{ij}\right)\frac{m_{hhr}}{m_{hhw}} - E_{hhj} \end{aligned} \tag{4.46}$$

and for light-holes

$$\begin{aligned} E_{a,ij} &= \left(h\nu - E_{ij}\right)\frac{m_{lhr}}{m_{cw}} + E_{ci} \\ E_{b,ij} &= -\left(h\nu - E_{ij}\right)\frac{m_{lhr}}{m_{lhw}} - E_{lhj} \end{aligned} \tag{4.47}$$

where $m_{hhr}$ and $m_{lhr}$, the reduced heavy-hole and light-hole well effective masses, are given by

$$\begin{aligned} m_{hhr} &= \frac{m_{cw} m_{hhw}}{m_{cw} + m_{hhw}} \\ m_{lhr} &= \frac{m_{cw} m_{lhw}}{m_{cw} + m_{lhw}} \end{aligned} \tag{4.48}$$

The CB subband to VB subband transition energy $E_{ij}$ is given by

$$\begin{aligned} E_{ij} &= E_g + E_{ci} + E_{hhj} \qquad \text{heavy-holes} \\ E_{ij} &= E_g + E_{ci} + E_{lhj} \qquad \text{light-holes} \end{aligned} \tag{4.49}$$

$H_{mod}$ is the modified Heaviside function defined as

$$H_{\text{mod}}\left(h\nu - E_{ij}\right) = \frac{1}{2} + \frac{1}{\pi}\tan^{-1}\left[\frac{\tau_{in}\left(h\nu - E_{ij}\right)}{h}\right] \tag{4.50}$$

$\tau_{in}$ is of the order of 0.1 ps. This function accounts for transition linebroadening. $C_{ij}$ is a spatial overlap factor given approximately by

$$C_{ij} \approx \begin{Bmatrix} 1, & i = j \\ 0, & i \neq j \end{Bmatrix} \tag{4.51}$$

$A_{ij}$ is an anisotropy factor that accounts for the polarisation of the electromagnetic field. Approximate anisotropy factors are given in Table 4.1.

*Table 4.1.* Quantum-well anisotropy factor

| Polarisation | Heavy-hole | Light-hole |
|---|---|---|
| TE | 1.5 | 0.5 |
| TM | 0 | 2.0 |

As before the gain coefficient $g'_m$ and absorption coefficient $g''_m$ can be obtained by simply replacing $f_c - f_v$ in (4.45) by $f_c[1-f_v]$ and $f_v[1-f_c]$ respectively. The gain coefficient of an unstrained quantum-well, as described above, exhibits strong polarisation dependency. This is because induced optical transitions from the CB to the heavy-hole band are much more favourable for TE polarised light (with its electric field parallel to the well layer) compared to TM polarised light (electric field normal to the well layer). Transitions from the CB to the light-hole band, which favour TM polarised light, are much less significant. While this effect is useful in laser design, it is not desirable in SOAs where low polarisation dependency is required. The polarisation dependency can be controlled by the introduction of strain in the quantum-well as discussed in Section 4.3.3.

EXAMPLE OF GAIN COEFFICIENT CALCULATION

In this example we calculate the gain spectrum of a 6.0 nm thick $In_{0.53}Ga_{0.47}As$ well surrounded by $In_{0.53}Ga_{0.47}As_{0.55}P_{0.45}$ barriers lattice matched to an InP substrate. The material has the following parameters: $E_{gapW}$ = 0.72 eV, $E_{gapB}$ = 0.96 eV, $E_p$ = 23.6 eV, $m_{cw}$ = 0.041$m_o$, $m_{cb}$ = 0.059$m_0$, $m_{vw}$ = 0.424$m_o$, $m_{vb}$ = 0.468$m_o$, $n_r$ = 3.22 and $T$ = 300 K. The bandgap discontinuities are given [13] by

$$\begin{aligned} \Delta E_c &= 0.4\Delta E_g \\ \Delta E_c &= 0.6\Delta E_g \end{aligned} \tag{4.52}$$

where $\Delta E_g$ is the difference between the barrier and well bandgap energies. Applying (4.41) to electrons and holes gives $N_c$ = 1, $N_{hh}$ = 2 and $N_{lh}$ = 1. Solving (4.40) for electrons and holes gives $E_{c0}$ = 48.9 meV, $E_{hh0}$ = 18.8

meV, $E_{hh1}$ = 74.0 meV and $E_{lh0}$ = 53.7 meV. The quasi-Fermi levels can be obtained from the numerical solution of (4.44). Typical plots of $g_m$ and $g'_m$

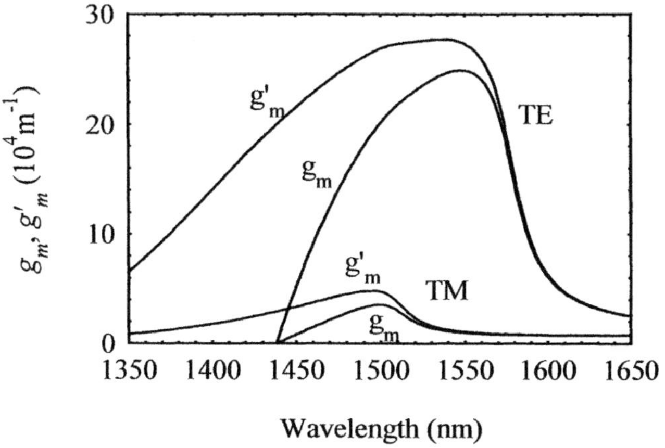


*Figure 4.9* Typical spectral plots of $g_m$ and $g'_m$ for a 6 nm thick unstrained InGaAsP-InGaAs quantum-well. The carrier density is 3.0 x $10^{24}$ m$^{-3}$.

spectra are shown in Fig. 4.9. Plots of TE $g_m$ spectra with carrier density as parameter are shown in Fig. 4.10. As in the case of a bulk semiconductor, analytic approximations to the gain coefficient can be used for fast gain calculations [13-15]

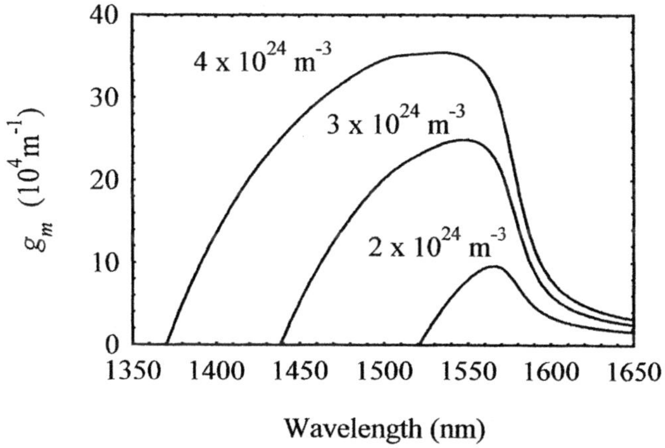


*Figure 4.10* Typical TE polarisation gain coefficient spectra for a 6 nm thick InGaAs-InGaAsP quantum-well. The parameter is carrier density.

Compared to bulk material the gain coefficient of an SQW is large. However the optical confinement factor of an SQW is small (typically 0.02 for an 8 nm well thickness) thereby offsetting the increased gain [16]. The optical confinement factor of an SQW depends on the well thickness and the refractive indices of the well and barrier regions. The main difference between an SQW and MQW material is the optical confinement factor. The confinement factor of MQW material is given approximately by $\Gamma_{MQW} = N_w \Gamma_{SQW}$ where $\Gamma_{QW}$ is the SQW confinement factor and $N_w$ the number of wells. Here it is assumed that all well are identical. The confinement factor can be further improved by use of the structures of Fig. 4.7.

### 4.2.3 Strained-layer quantum-well SOAs

The polarisation sensitivity of MQW SOAs can be greatly reduced by the introduction of strain in the wells [5,7-8,17-19]. Strain can be introduced by creating a lattice mismatch between the well and the adjacent barrier layers. In a lattice-matched quantum-well the heavy- and light-hole VBs are degenerate (i.e. they occupy the same energy and momentum space). The effect of strain is to reduce this degeneracy. Strain also changes electron and hole effective masses.

There are two type of strain: Compressive and tensile. The effects of strain on the energy band diagram of a quantum-well are shown in Fig. 4.11. The CB band edge shift $\delta E_c$, heavy-hole and light-hole band edge shifts $\delta E_{hh}$ and $\delta E_{lh}$ are given by

$$\begin{aligned}
\delta E_c &= -a_c(2\varepsilon_{xx} + \varepsilon_{zz}) \\
\delta E_{hh} &= a_v(2\varepsilon_{xx} + \varepsilon_{zz}) + b(\varepsilon_{xx} - \varepsilon_{zz}) \\
\delta E_{lh} &= a_v(2\varepsilon_{xx} + \varepsilon_{zz}) - b(\varepsilon_{xx} - \varepsilon_{zz})
\end{aligned} \tag{4.53}$$

where

$$\begin{aligned}
\varepsilon_{xx} &= \frac{a_0 - a}{a} \\
\varepsilon_{zz} &= -2\frac{C_{12}}{C_{11}}\varepsilon_{xx}
\end{aligned} \tag{4.54}$$

$a_c$, $a_v$ and $b$ are the CB, VB and shear deformation potentials respectively. $a$ and $a_0$ are the well and barrier lattice constants respectively. $C_{11}$ and $C_{12}$ are shear elastic coefficients respectively. The values of these parameters and the electron and hole effective masses depend on the well and barrier composition. Typical values of these parameters for InAs and GaAs are listed in Table 4.2, from which the appropriate values for an $In_{1-x}Ga_xAs$ well (depending on the Gallium composition $x$) can be extrapolated.

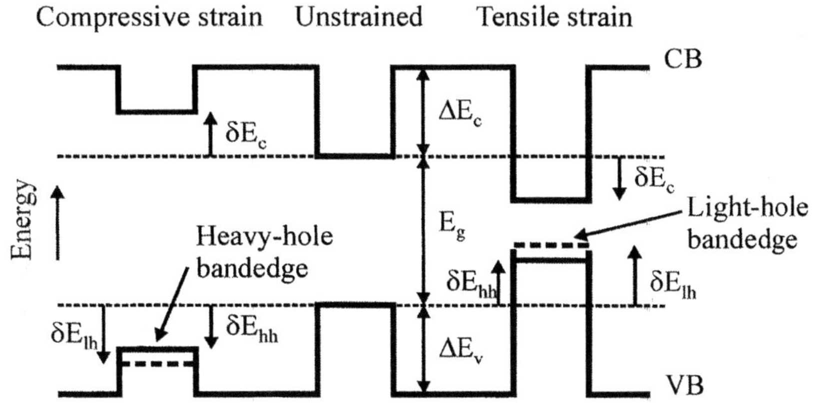


*Figure 4.11.* Band edge profile for compressive strained, unstrained and tensile strained quantum-wells.

*Table 4.2.* Some material parameters for InAs and GaAs [5].

| Parameter | Symbol | InAs | GaAs |
|---|---|---|---|
| Lattice constant | $a$ | 6.0584 | 5.6533 |
| CB deformation potential (eV) | $a_c$ | -5.08 | -7.17 |
| VB deformation potential (eV) | $a_v$ | 1.00 | 1.16 |
| Shear deformation potential (eV) | $b$ | -1.8 | -1.7 |
| Elastic coefficient ($10^{15}$ dyne/m$^2$) | $C_{11}$ | 8.329 | 11.879 |
| Elastic coefficient ($10^{15}$ dyne/m$^2$) | $C_{12}$ | 4.526 | 5.376 |

In a compressively strained quantum-well the heavy-hole band edge is closer to the CB edge than the light-hole band edge. This leads to an enhancement of the TE gain at the expense of the TM mode gain. If enough tensile strain is introduced in a quantum-well then the TM gain can exceed the TE gain. It can be difficult to determine the well or barrier composition required to equalise the TE and TM gains and so achieve polarisation insensitive gain. The use of tensile strained quantum-wells can lead to an increase in the transparency carrier density and thereby an increase in the noise figure [20].

Polarisation insensitive MQW SOAs can also be fabricated using tensile and compressively strained quantum-wells in a single active layer [21-22]. In

composition or well width) in order to equalise the TE and TM mode gain coefficients. Also the ratio of the number of both well types can be adjusted to compensate for the different TE and TM optical confinement factors. Polarisation insensitive MQW SOAs have also been fabricated where the strain is introduced in the barrier layers of the quantum-well active layer [23].

The area of quantum-well SOA design and fabrication is undergoing rapid research and development. Many quantum-well configurations are possible that deliver improvements in particular SOA characteristics and performance [24-27].

### 4.2.2 Comparison between bulk and quantum-well SOAs

GAIN BANDWIDTH

The 3 dB bandwidth of the gain coefficient of a quantum-well, at the carrier densities found in SOAs, is greater than that for a bulk semiconductor. This is primarily due to the different density of states function the form of which are plotted in Fig. 4.12 for comparison.

In bulk semiconductors the density of states is a continuous function of energy. This means that the gain coefficient is strongly dependent on the photon energy (see Fig. 4.3). On the other hand, the density of states for a quantum-well has a staircase type dependency on energy. This reduced dependence on energy leads to a flatter gain coefficient spectrum. This implies that quantum-well SOAs tend to have superior gain bandwidths compared to bulk devices.

POLARISATION DEPENDENCY

The main cause of polarisation dependency in bulk SOAs is the difference between the TE and TM confinement factors. Strained quantum-well SOAs can be used to offset this difference and so obtain low polarisation sensitivity.

SATURATION OUTPUT POWER

The saturation output power of an SOA is inversely proportional to the differential gain coefficient $a_l$ (Chapter 3). The value of this coefficient is generally smaller in quantum-well SOAs compared to bulk SOAs leading to improved saturation characteristics. However the optical confinement factor of an MQW active region having the same overall thickness as a bulk active region is lower, leading to a lower amplifier gain. MQW SOAs generally deliver higher gain at relatively low carrier density, which tends to offset the effect of the lower confinement factor. Simply increasing the device length can also reduce the effect of the lower confinement factor. The confinement factor can also be increased by the use of GRIN or modified MQW structures (Fig. 4.7).

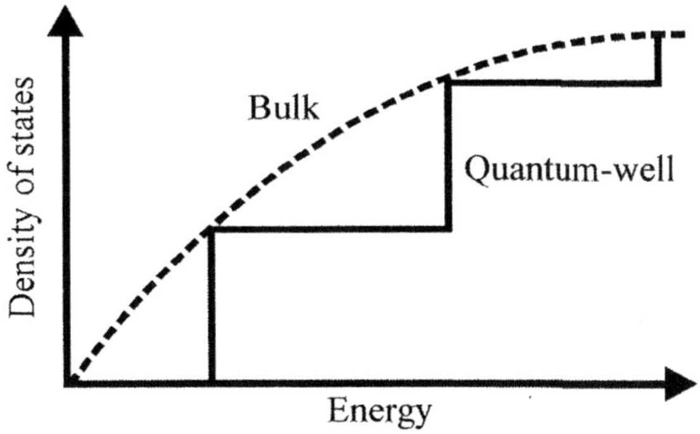


*Figure 4.12* Form of the density of states functions for bulk and quantum-well semiconductors. In practice the edges of the quantum-well density of states function are rounded due to the transition linebroadening.

### 4.2.4 Carrier recombination mechanisms in quantum-wells

The recombination mechanisms in quantum-wells are similar to those occurring in bulk semiconductors and in general can modelled by the same set of equations given in Section 4.1.3. The actual values of the recombination coefficients used in those equations depend on the type of quantum-well [28-29]. The most important recombination coefficient $B_{rad}$ is typically of the order of 1 x $10^{-16}$ $m^3s^{-1}$.

## 4.4 LOSS MECHANISMS

In addition to the intrinsic material absorption, light propagating in an SOA is subject to a number of additional loss mechanisms. The total loss coefficient of an SOA active region can by modelled [4] by

$$\alpha(n) = K_0 + \Gamma K_1 n \tag{4.55}$$

where $K_0$ and $K_1$ are the carrier independent and carrier dependent absorption loss coefficients respectively. $K_0$ represents the intrinsic material and waveguide losses. $K_l$ is mainly due to intervalence band absorption and carrier dependent scattering losses. The net gain coefficient $g$ of an SOA is then given by

$$g = \Gamma g_m - \alpha(n) \tag{4.56}$$

# REFERENCES


1. A. Yariv, *Quantum Electronics*, John Wiley & Sons (1989).
2. A. Yariv, *Optical Electronics*, HRW International (1985).
3. G.P. Agrawal and N.K. Dutta, *Long-Wavelength Semiconductor Lasers*, Van Nostrand Reinhold (1986).
4. Y. Suematsu and A.R. Adams, *Handbook of Semiconductor lasers and Photonic Integrated Circuits*, London, Chapman and Hall (1994).
5. S. Lien Chuang, *Physics of Optoelectronic Devices*, New York, Wiley Interscience, (1995).
6. S. Adachi, *Physical Properties of III-IV Compounds*, New York: Chichester, John Wiley and Sons (1992).
7. J.P. Loehr, *Physics of Strained Quantum-Well Lasers*, Kluwer Academic Publishers (1998).
8. P.S. Zory, *Quantum Well Lasers*, Academic Press (1993).
9. N.G. Nilsson, Empirical approximations for the Fermi energy of a semiconductor with parabolic bands, *Appl. Phys. Lett.*, **33**, 653-654 (1978).
10. R. Olshansky, C.A. Su, J. Manning and W. Powazinik, Measurement of radiative and nonradiative recombination rates in InGaAsP and AlGaAs light sources, *IEEE J. Quantum Electron.*, **20**, 838-854 (1984).
11. M. Kot and K. Zdansky, Measurement of radiative and nonradiative recombination rate in InGaAsP-InP LED's, *IEEE J. Quantum Electron.*, **28**, 1746-1750 (1992).
12. M. Asada, A. Kameyama and Y. Suematsu, Gain and intervalence band absorption in quantum-well lasers, *IEEE J. Quantum Electron.*, **20**, 745-753 (1984).
13. T. Makino, Analytical formulas for the optical gain of quantum wells, Makino, *IEEE J. Quantum Electron.*, **32**, 493-501 (1996).
14. S. Balle, Simple analytical approximations for the gain and refractive index spectra in quantum-well lasers, *Phys. Rev. A*, **57**, 1304-1312 (1998).
15. T.-A. Ma, Z.-M. Li, T. Makino and M.S. Wartak, Approximate optical gain formulas for 1.55 μm strained quaternary quantum-well lasers, *IEEE J. Quantum Electron.*, **31**, 29-34 (1995).
16. J. Kraus and P.P. Deimel, Full-width-half-maximum and confinement of optical modes in multiple-quantum-well laser structures, *J. Lightwave Technol.*, **11**, 1802-1805 (1993).
17. E.P. O'Reilly and A.R. Adams, Band-structure engineering in strained semiconductor lasers, *IEEE J. Quantum Electron.*, **30**, 366-379 (1994).
18. P.J.A. Thijs, L.F. Tiemeijer, J.J.M. Binsma and T. van Dongen, Progress in long-wavelength strained-layer InGaAs(P) quantum well semiconductor lasers and amplifiers, *IEEE J. Quantum Electron.*, 477-499 (1994).
19. Li. Zhan-Ming, M. Dion, Z. Yao, J. Wang, M. Davies and S.P. McAlister, An approximate k/spl middot/p theory for optical gain of strained InGaAsP quantum-well lasers, *IEEE J. Quantum Electron.*, **30,** 538 -546 (1994).

20. M. Joma, H. Horikawa, C.Q. Xiu, K. Yamada, Y. Katoh and T. Kamijoh, Polarisation insensitive semiconductor laser amplifiers with tensile strained InGaAsP/InGaAsP multiple quantum well structure, *Appl. Phys. Lett.*, **62**, 121-122 (1993).
21. M.A. Newkirk, B. I. Miller, U. Koren, M. G. Young, M. Chien, R. M. Jopson, and C. A. Burrus, 1.5 μm multiquantum-well semiconductor optical amplifier with tensile and compressively strained wells for polarisation-independent gain, *IEEE Photon. Technol. Lett.*, **4**, 406-408 (1993).
22. M. Silver, A.F. Phillips, A.R. Adams, P.D. Greene and A.J. Collar, Design and ASE characteristics of 1550-nm polarization-insensitive semiconductor optical amplifiers containing tensile and compressive wells, *IEEE J. of Quantum Electron.*, **36**, 118-122 (2000).
23. K. Magari, M. Otomoto and Y. Noguchi, Polarisation-insensitive optical amplifier with tensile-strained-barrier MQW structure, *IEEE J. Quantum. Electron.*, **30**, 695-702 (1994).
24. F. Seiferth, F.G. Johnson, S.A. Merritt, S. Fox, R.D. Whaley, Y.J. Chen, M. Dagenais and D.R. Stone, Polarization insensitive 1.55 μm optical amplifier with GaAs delta-strained $Ga_{0.47}In_{0.53}As$ quantum wells, *IEEE Photon. Technol. Lett.*, **9**, 1340 -1342 (1997).
25. C. Yong-Sang and C. Woo-Young, Analysis and optimization of polarization-insensitive semiconductor optical amplifiers with delta-strained quantum wells, *IEEE J. Quantum Electron.*, **37**, 574 -579 (2001).
26. R. Zhang and P.P. Ruden, 1.3 μm polarisation insensitive optical amplifier structure based on coupled quantum wells, *IEEE J. Quantum Electron.*, **35**, 1509-1514 (1999).
27. P. Granestrand, Proposal for an improved quantum-well laser amplifier with wide wavelength operation, *IEEE Photon. Technol. Lett.*, **8**, 203-205 (1996).
28. N. Storkfelt, M. Yamaguchi, B. Mikkelsen and K.E. Stubkjaer, Recombination constants and alpha-factor in 1.5μm MQW optical amplifiers taking carrier overflow into account, *Elec. Lett.*, **28**, 1774-1776, (1992).
29. J.M. Pikal, C.S. Menoni, H. Temkin, P. Thiagarajan and G.Y. Robinson, Carrier lifetime and recombination in long-wavelength quantum-well lasers, *IEEE J. Select. Topics in Quantum Electron.*, **5**, 613-618 (1999).

Chapter 5

# MODELLING

Models of SOA steady-state and dynamic behaviour are important tools that allow the SOA designer to develop optimised devices (typical desirable properties are listed in Table 2.1). They also allow the applications engineer to predict how an SOA or cascade of SOAs behaves in a particular application. In this chapter we concentrate on specific models and use them to focus on the main characteristics of SOAs.

## 5.1 STEADY-STATE MODEL

Many SOA steady-state models, of varying degrees of complexity are available [1-8]. Some models are amenable to analytical solution while others require numerical solution. The analytical techniques are usually only applicable over narrow operating regimes and tend to be inaccurate, however they are useful in obtaining insights into SOA behaviour. Numerical techniques are usually more complex but make fewer assumptions and are often applicable over a wide range of operating regimes. With the advent of fast personal computers, numerical techniques are beginning to supersede analytical techniques. In this section we use a steady-state model, developed by the author [8], to explore the main characteristics of a bulk homogeneous buried ridge stripe SOA [9]. Other types of SOA can be modelled in a similar fashion.

The main purpose of an SOA model is to relate the internal variables of the amplifier to measurable external variables such as the output signal power, saturation output power and amplified spontaneous emission (ASE) spectrum. This aids the design and optimisation of an SOA for a given application.

The particular SOA under consideration in this section is a 1.55 μm $InP/In_{1-x}Ga_xAs_yP_{1-y}$ homogeneous buried ridge stripe device (Fig. 3.9). $y$ and $x$ are the molar fractions of Arsenide and Gallium respectively in the undoped active region. Lattice matching is assumed for which $x = 0.47y$. The device structure consists of central active region of width $W$, thickness $d$ and length $L_c$. The active region narrows linearly as a lateral taper of width $W$ at the central region to zero width at each end. Each taper has length $L_t$. The tapers reduce the optical confinement factor from its maximum value $\Gamma$ at the central active region to zero at the amplifier ends. This causes the guided mode lateral profile to expand which improves the input and output coupling efficiencies to single mode optical fibre.

As a first approximation the SOA is modelled as a device of mean length $L$ given by

$$L = L_c + L_t \tag{5.1}$$

with identical geometrical properties and optical confinement factor as the central active region. Geometrical and material parameters for the device under consideration are given in Table 5.1. The material gain coefficient is calculated using the full model given in Chapter 4. The bandgap energy can be expressed by

$$E_g(n) = E_{g0} - eK_g n^{1/3} \tag{5.2}$$

where $K_g$ is the bandgap shrinkage coefficient [10]. The value of $K_g$ used in the model is taken to be slightly less than the value for $In_{0.15}Ga_{0.85}As$. The main effect of bandgap shrinkage is to shift the peak of the gain and spontaneous emission spectra towards longer wavelengths. $E_{g0}$, the bandgap energy with no injected carriers, is given by the quadratic approximation [10]

$$E_{g0} = e\left(a + by + cy^2\right) \tag{5.3}$$

where $a$, $b$ and $c$ are the quadratic coefficients and $e$ the electronic charge.

### 5.1.1 Travelling-wave equations for the signal field

In the model $N_s$ signals are injected into the amplifier with optical frequencies $\nu_k$ ($k = 1...N_s$) and power $Pin_k$ before coupling loss. The signals travel through the amplifier, aided by the embedded waveguide, and exit at the opposite facet. In the following analysis it is assumed that transverse variations (i.e. normal to the propagation direction) in the photon rates and carrier density are negligible. This is a valid assumption because most SOAs have a narrow active region. In the model the left (input) and right (output) facets have power reflectivities $R_1$ and $R_2$ respectively. In the amplifier the spatially varying component of the field due to each signal can be decomposed into two complex travelling-waves $Es_k^+$ and $Es_k^-$, propagating in the positive and negative $z$ directions respectively. $z$ lies along the amplifier axis with its origin at the input facet. It is assumed that the modulus squared of the amplitude of a travelling-wave is equal to the photon rate ($s^{-1}$) of the wave in that direction, so

$$
\begin{aligned}
Ns_k^+ &= \left|Es_k^+\right|^2 \\
Ns_k^- &= \left|Es_k^-\right|^2
\end{aligned}
\tag{5.4}
$$

*Table 5.1.* SOA geometrical and material parameters used in the steady-state model [8].

| Symbol | Parameter | Value |
|---|---|---|
| $y$ | Molar fraction of Arsenide in the active region | 0.892 |
| $L_c$ | Central active region length. | 600 μm |
| $L_t$ | Tapered active region length. | 100 μm |
| $d$ | Active region thickness. | 0.4 μm |
| $W$ | Central active region width. | 0.4 μm |
| $\Gamma$ | Optical confinement factor. | 0.45 |
| $Kg$ | Bandgap shrinkage coefficient. | $0.9 \times 10^{-10}$ eVm |
| $n_{eff}$ | Effective index at zero carrier density. | 3.22 |
| $\eta_{in}$ | Input coupling loss. | 3.0 dB |
| $\eta_{out}$ | Output coupling loss. | 3.0 dB |
| $R_1$ | Input facet reflectivity. | $5.0 \times 10^{-5}$ |
| $R_2$ | Output facet reflectivity. | $5.0 \times 10^{-5}$ |
| $K_0$ | Carrier independent absorption loss coefficient. | 6200 $m^{-1}$ |
| $K_1$ | Carrier dependent absorption loss coefficient. | $7500 \times 10^{-24}$ $m^2$ |
| $A_{rad}$ | Linear radiative recombination coefficient. | $1.0 \times 10^{7}$ $s^{-1}$ |
| $B_{rad}$ | Bimolecular radiative recombination coefficient. | $5.6 \times 10^{-16}$ $m^3s^{-1}$ |
| $A_{nrad}$ | Linear nonradiative recombination coefficient due to traps. | $3.5 \times 10^{8}$ $s^{-1}$ |
| $B_{nrad}$ | Bimolecular nonradiative recombination coefficient. | $0.0 \times 10^{-16}$ $m^3s^{-1}$ |
| $C_{aug}$ | Auger recombination coefficient. | $3.0 \times 10^{-41}$ $m^6s^{-1}$ |
| $D_{leak}$ | Leakage recombination coefficient. | $0 \times 10^{48}$ $m^{13.5}s^{-1}$ |
| $a$ | Bandgap energy quadratic coefficient. | 1.35 |
| $b$ | Bandgap energy quadratic coefficient. | -0.775 |
| $c$ | Bandgap energy quadratic coefficient. | 0.149 |
| $m_e$ | Effective mass of electron in the CB. | $4.10 \times 10^{-32}$ kg |
| $m_{hh}$ | Effective mass of a heavy hole in the VB. | $4.19 \times 10^{-31}$ kg |
| $m_{lh}$ | Effective mass of a light hole in the VB. | $5.06 \times 10^{-32}$ kg |

The light wave representing the signal must be treated coherently since its transmission through the amplifier depends on its frequency and phase when reflecting facets are present. $Es_k^+$ and $Es_k^-$ obey the complex travelling-wave equations

$$\frac{dEs_k^+(z)}{dz} = \left\{-j\beta_k + \frac{1}{2}\left[\Gamma g_m(\nu_k, n) - \alpha\right]\right\} Es_k^+(z)$$
$$\frac{dEs_k^-(z)}{dz} = \left\{j\beta_k - \frac{1}{2}\left[\Gamma g_m(\nu_k, n) - \alpha\right]\right\} Es_k^-(z) \tag{5.5}$$

where $j = \sqrt{-1}$ and $\alpha$ is the material loss coefficient . (5.5) is subject to the boundary conditions

$$Es_k^+(0) = (1 - r_1)Ein_k + r_1 Es_k^-(0)$$
$$Es_k^-(L) = r_2 Es_k^+(L) \tag{5.6}$$

where the k-th input signal field to the left of the input facet is

$$Ein_k = \sqrt{\frac{\eta_{in} Pin_k}{h\nu_k}} \tag{5.7}$$

The k-th output field to the right of the output facet is

$$Eout_k = (1 - r_2)Es_k^+(L) \tag{5.8}$$

The k-th output signal power after coupling loss is

$$Pout_k = h\nu_k \eta_{out} \left|Eout_k\right|^2 \tag{5.9}$$

$\eta_{in}$ and $\eta_{out}$ are the input and output coupling efficiencies respectively. The amplitude reflectivity coefficients are

$$r_1 = \sqrt{R_1}$$
$$r_2 = \sqrt{R_2} \tag{5.10}$$

The signal propagation coefficient is given by

$$\beta_k = \frac{2\pi n_{eff} \nu_k}{c} \tag{5.11}$$

where $n_{eff}$ is the effective index of the amplifier waveguide. The actual value of $n_{eff}$ determines the position of the ripples in the amplifier gain and spontaneous emission spectra, the exact location of which is not an important operational parameter in SOAs with low facet reflectivities.

### 5.1.2 Travelling-wave equations for the spontaneous emission

The amplification of the signal also depends on the amount of spontaneously emitted noise generated by the amplifier. This is because the noise power takes part in draining the available carrier population and helps saturate the gain. However, it is not necessary to treat the spontaneous emission as a coherent signal since it distributes itself continuously over a relatively wide band of wavelengths with random phases between adjacent wavelength components. When reflecting facets are present the spontaneously emitted noise will show the presence of longitudinal cavity modes. For this reason it can be assumed that noise photons only exist at discrete frequencies corresponding to integer multiples of cavity resonances. These frequencies are given by

$$\nu_j = \nu_c + \Delta\nu_c + jK_m \Delta\nu_m \quad ; j = 0 \ldots N_m - 1 \tag{5.12}$$

where the cut-off frequency is given by

$$\nu_c = \frac{E_g}{h} \tag{5.13}$$

$\Delta\nu_c$ is a frequency offset used to let $\nu_0$ correspond to a resonance. $K_m$ and $N_m$ are positive integers. The values of $K_m$ and $N_m$ chosen depend on the gain bandwidth of the SOA and accuracy required from the numerical solution of the model equations. The longitudinal mode frequency spacing is

$$\Delta\nu_m = \frac{c}{2n_{eff} L} \tag{5.14}$$

This technique can be applied to both resonant and near-travelling-wave SOAs and greatly reduces computation time. It can be shown that averaging the coherent signal over two adjacent cavity resonances is identical to treating the signal coherently in terms of travelling-wave power (or photon rate) equations. It is sufficient to describe the spontaneous emission in terms of power, while signals must be treated in terms of waves with definite amplitude and phase.

The spontaneous emission photon rates $N_j^+$ and $N_j^-$ ($s^{-1}$) with a particular polarisation (TE or TM) in a frequency spacing $K_m \Delta \nu_m$ centred on frequency $\nu_j$ travelling in the positive and negative directions along the amplifier axis respectively obey the travelling-wave equations

$$\begin{aligned}\frac{dN_j^+(z)}{dz} &= \left[\Gamma g_m(\nu_j,n) - \alpha_s(n)\right]N_j^+ + R_{sp}(\nu_j,n)\\ \frac{dN_j^-(z)}{dz} &= -\left[\Gamma g_m(\nu_j,n) - \alpha_s(n)\right]N_j^- - R_{sp}(\nu_j,n)\end{aligned} \tag{5.15}$$

subject to the boundary conditions

$$\begin{aligned}N_j^+(0) &= R_1 N_j^-(0)\\ N_j^-(L) &= R_2 N_j^+(L)\end{aligned} \tag{5.16}$$

$R_{sp}(\nu_j,n)$ represents the spontaneous emission coupled into $N_j^+$ or $N_j^-$. An expression for $R_{sp}$ can be derived by a comparison between the noise output from an ideal amplifier obtained using (5.15) with the quantum mechanically derived expression [11]. An ideal amplifier has no gain saturation (which implies a constant carrier density throughout the amplifier), material gain coefficient $g_m = g'_m\ (> 0)$ and zero loss coefficient, facet reflectivities and coupling losses. In this case $N_j^+$ is obtained from the solution to

$$\frac{dN_j^+}{dz} = \Gamma g'_m(\nu_j,n)N_j^+ + R_{sp}(\nu_j,n) \tag{5.17}$$

The solution of (5.17) in this ideal case gives an output noise power at a single polarisation into a frequency band $K_m \Delta \nu_m$ centred on $\nu_j$ of

$$P_{out} = \frac{G-1}{\Gamma g'_m(\nu_j, n)} h\nu_j R_{sp}(\nu_j, n) \tag{5.18}$$

where $G$ is the single-pass gain at $\nu_j$. The equivalent quantum mechanical expression [12] is

$$N_{out} = (G-1) h\nu_j K_m \Delta\nu_m \tag{5.19}$$

Equating (5.18) with (5.19) gives

$$R_{sp}(\nu_j, n) = \Gamma g'_m(\nu_j, n) K_m \Delta\nu_m \tag{5.20}$$

The travelling-wave power equations describing $N_j^+$ and $N_j^-$ assume that all the spontaneous photons in spacing $K_m \Delta\nu_m$ are at resonance frequencies. In a real device the injected spontaneous photons, originating from $R_{sp}$, are uniformly spread over $K_m \Delta\nu_m$. The noise is filtered by the amplifier cavity. To account for this $N_j^+$ and $N_j^-$ are multiplied by a normalization factor $K_j$ which is derived as follows.

The signal gain for frequencies within spacing $\Delta\nu_m$ around $\nu_j$, from (2.25) is

$$G(\nu) = \frac{(1-R_1)(1-R_2)G_s}{\left(1-\sqrt{R_1R_2}G_S\right)^2 + 4G_S\sqrt{R_1R_2}Sin^2\phi} \tag{5.21}$$

where the single-pass phase shift is

$$\phi = \frac{2\pi n_{eff}\nu}{c} \tag{5.22}$$

and single-pass gain is

$$G_s = \exp\left\{\int_0^L \left[\Gamma g_m(\nu_j, n) - \alpha(n)\right] dz\right\} \tag{5.23}$$

At resonance the signal gain is

$$G(\nu_j) = \frac{(1-R_1)(1-R_2)G_s}{\left(1-\sqrt{R_1R_2}G_s\right)^2} \tag{5.24}$$

Let the amplifier have a noise input spectral density $\sigma_{in}$ (photons/s/Hz) distributed uniformly over $\Delta\nu_m$ centred at $\nu_j$. The total output noise (photons/s) in $\Delta\nu_m$ is then

$$\begin{aligned} N_{out} &= \sigma_{in} \int_{\nu=\nu_j-\Delta\nu_m/2}^{\nu=\nu_j+\Delta\nu_m/2} G(\nu)d\nu \\ &= \sigma_{in}\Delta\nu_m \frac{1}{\pi}\int_0^{\pi} G(\phi)d\phi \end{aligned} \tag{5.25}$$

If the input noise power were concentrated at $\nu_j$ (resonance), then the output noise photon rate would be

$$N_{j,out} = K_j \sigma_{in} G(\nu_j)\Delta\nu_m \tag{5.26}$$

The factor $K_j$ is included in (5.26) to equate $N_{out}$ with $N_{j,out}$ so

$$\begin{aligned} K_j &= \frac{\int_0^{\pi} G(\phi)d\phi}{\pi G(\nu_j)} \\ &= \frac{1}{\sqrt{1+\gamma^2}} \end{aligned} \tag{5.27}$$

where

$$\gamma = \frac{4G_s(\nu_j)\sqrt{R_1R_2}}{\left[1-\sqrt{R_1R_2}G_S(\nu_j)\right]^2} \tag{5.28}$$

$K_j$ is equal to unity for zero facet reflectivities.

### 5.1.3 Carrier density rate equation

The amplifier material gain and spontaneous emission originate from carriers injected into the active region by the bias current. The carrier density ($m^{-3}$) at position $z$ in the amplifier obeys the rate equation

$$\begin{aligned}\frac{dn(z)}{dt} = {} & \frac{I}{edLW} - R(n) \\ & - \frac{\Gamma}{dW}\left\{\sum_{k=1}^{N_s} g_m(\nu_k, n)\left[Ns_k^+(z) + Ns_k^-(z)\right]\right\} \\ & - \frac{2\Gamma}{dW}\left\{\sum_{j=0}^{N_m-1} g_m(\nu_j, n)K_j\left[N_j^+(z) + N_j^-(z)\right]\right\}\end{aligned} \tag{5.29}$$

where $I$ is the amplifier bias current. In (5.29) all of the bias current is assumed to pass through the active region and not the surrounding InP regions. The bias current is assumed to have a uniform distribution across the active region width.

The first term on the RHS of (5.29) represents the addition of carriers to the active region from the bias current. These injected carriers are then depleted by various mechanisms occurring within the amplifier. The second term is due to radiative and nonradiative recombination mechanisms discussed in Chapter 4. It is assumed that carrier leakage from the active region into surrounding InP regions is negligible [13]. The third and fourth terms on the RHS of (5.29) represent radiative recombination of carriers due to the amplified signal and ASE. $\Gamma$ is included as only this fraction of amplified photons resides in the active region. The factor of two in (5.29) accounts for the fact that spontaneously emitted photons can exist in one of two mutually orthogonal polarisations (TE or TM). In the model the SOA is assumed to be polarisation independent. Polarization dependence can be included by the use of different TE and TM confinement factors.

### 5.1.4 Numerical algorithm

As the SOA model equations cannot be solved analytically, a numerical solution is required. In the numerical model the amplifier is split into a number of sections labelled from $i = 1$ to $N_z$ as shown in Fig. 5.1. The signal fields and spontaneous emission photon rates are estimated at the section interfaces. In evaluating the RHS of (5.29) $Q(i)$ for the i-th section, the signal and noise photon rates used are given by the mean value of those quantities at the section boundaries. In the steady-state $Q(i)$ is zero.

To predict the steady-state characteristics an algorithm is used that adjusts the carrier density so the value of *Q(i)* throughout the amplifier approaches zero. A flow chart of the algorithm is shown in Fig. 5.2. The first step in the algorithm is to initialise the signal fields and spontaneous emission photon rates to zero. The initial carrier density is obtained from the solution of (5.29) with all fields set to zero. The iteration then begins.

Firstly, the coefficients of the travelling-wave equations are computed. Next the signal fields and noise photon densities are estimated using finite difference solutions of (5.5) and (5.15) [14].

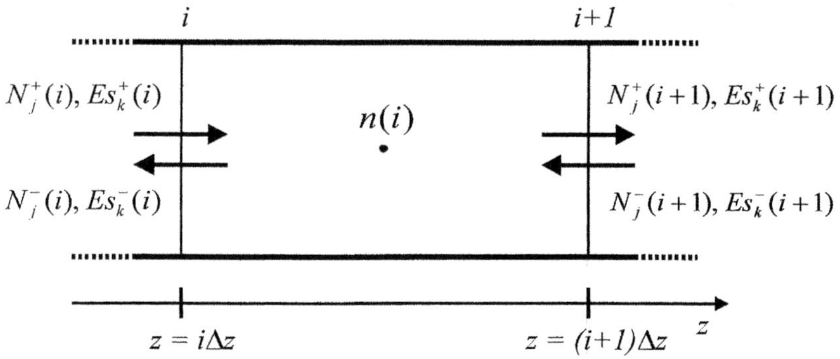


*Figure 5.1.* *i*-th section of the SOA model. The signal fields and spontaneous emission photon rated are estimated at the section boundaries. The carrier density is estimated at the centre of the section (after [11] with the permission of the IEEE).

Forward differences are used for positive travelling-waves and backward differences for negative travelling-waves. If $X^+$ and $X^-$ are forward and backward travelling-waves then for the i-th section the spatial derivatives are approximated by

$$\begin{aligned} \frac{dX^+}{dz} &= \frac{X_i^+ - X_{i-1}^+}{\Delta z} \\ \frac{dX^-}{dz} &= \frac{X_{i+1}^- - X_i^-}{\Delta z} \end{aligned} \qquad (5.30)$$

where $\Delta z = L/N_z$ is the length of a single section. *Q(i)* is then calculated for each section. If *Q(i)* is positive then the carrier density is too low, in which case the new value of carrier density for the section is increased by a factor *1* + *W(i)*, where *W(i)* is a weight less than unity. An initial value for *W(i)* of 0.1 was found to give good convergence. If *Q(i)* is negative then the carrier density is too high, in which case the new value of carrier density for the section is decreased by a factor $1 + W(i)$. If the sign of $Q(i)$ differs from the previous iteration, $W(i)$ is halved. This process enables convergence towards the correct value of carrier density by using smaller carrier density increments. The iteration continues until the percentage change in the signal fields, noise photon rates and carrier density throughout the SOA between successive iterations is less than the desired tolerance. When the iteration stops, the output spontaneous emission power spectral density is computed (Section 5.1.5) and parameters such as signal gain, noise figure and output spontaneous noise power are calculated. The algorithm shows good convergence and stability over a wide range of operating conditions.

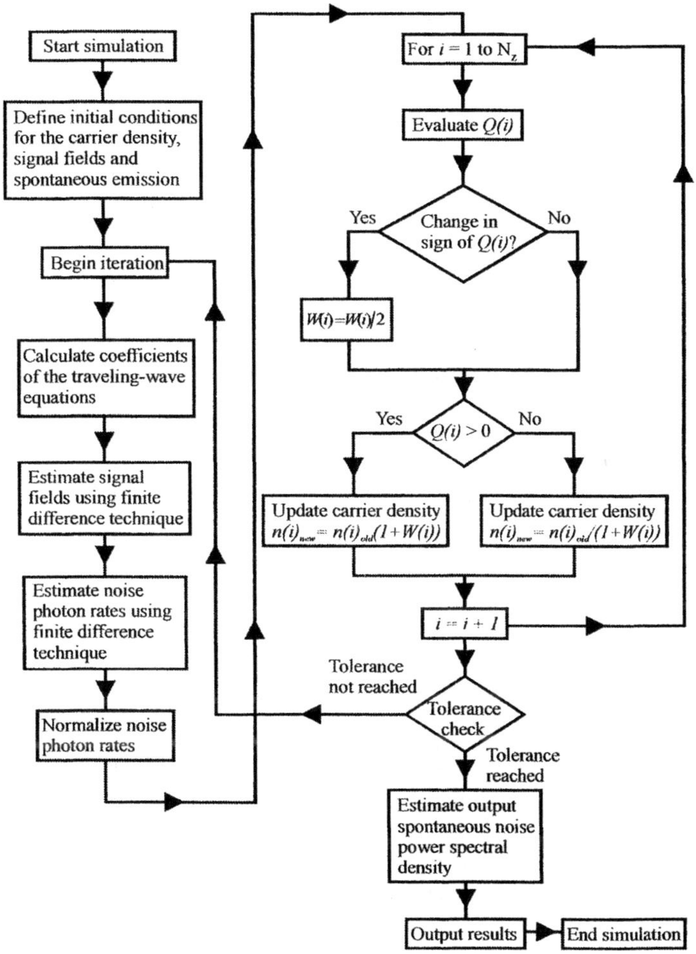


*Figure 5.2.* Flow chart of the SOA steady-state model algorithm. The algorithm updates the carrier density ($n(i)_{new}$) throughout the amplifier such that $Q(i)$ (the RHS of (5.29)) approaches zero (after [8] with the permission of the IEEE).

### 5.1.5 Output spontaneous emission power spectral density

The average output noise photon rate spectral density (photons/s/Hz) after the coupling loss over both polarisations and bandwidth $K_m \Delta \nu_m$ centred on $\nu_j$ is

$$\sigma_N(\nu_j) = 2\eta_{out}(1 - R_2)K_j N_j^+(L)/(K_m \Delta \nu_m), \quad j = 0 \ldots N_m - 1. \quad (5.31)$$

The average output noise photon rate spectral density $\sigma_N(\nu_{res,j})$ centred on the cavity resonance frequencies $\nu_{res,j}$ is interpolated using a cubic spline fit to the set of $\sigma_N(\nu_j)$ points [14]. The amplifier resonant frequencies are given by

$$\nu_{res,j} = \nu_c + \Delta \nu_c + j\Delta \nu_m, \quad j = 0 \ldots K_m(N_m - 1) \quad (5.32)$$

The anti-resonant frequencies are given by

$$\nu_{ares,j} = \nu_{res,j} - \Delta \nu_m/2, \quad j = 1 \ldots K_m(N_m - 1) \quad (5.33)$$

The SOA gains at the resonant and anti-resonant frequencies are

$$\begin{aligned} G_{res}(\nu_{res,j}) &= \frac{(1 - R_1)(1 - R_2)G_s(\nu_{res,j})}{\left[1 - \sqrt{R_1 R_2} G_S(\nu_{res,j})\right]^2} \\ G_{ares}(\nu_{ares,j}) &= \frac{(1 - R_1)(1 - R_2)G_s(\nu_{ares,j})}{\left[1 + \sqrt{R_1 R_2} G_S(\nu_{ares,j})\right]^2} \end{aligned} \quad (5.34)$$

Trapezoidal integration can be used to evaluate the single-pass gain $G_s$. The amplifier gain profile $T(\nu)$ can be obtained by fitting a cubic spline to the resonant and anti-resonant gain points. The output spontaneous emission noise power spectral density $\sigma_{ASE}(\nu)$ (*Watts/Hz*) is determined by multiplying $\sigma_N(\nu_{res,j})$ by the photon energy $h\nu$ and the ratio of the gain at that frequency $T(\nu)$ to the average gain of the closest longitudinal mode $j$ to $\nu$, i.e.,

$$\sigma_{ASE}(\nu) = \frac{h\nu\sigma_N(\nu_{res,j})T(\nu)}{\frac{1}{\Delta\nu_m}\int\limits_{\nu_{ares,j}}^{\nu_{ares,j+1}} T(\nu')d\nu'} \tag{5.35}$$

The integral in the denominator of (5.35) is the average gain of the j-th longitudinal mode. From (5.35) the output spontaneous emission power into a frequency (resolution) bandwidth $\Delta\nu$ centred on $\nu$ is

$$P_{ASE}(\nu) = \int\limits_{\nu-\Delta\nu/2}^{\nu+\Delta\nu/2} \sigma_{ASE}(\nu')d\nu' \tag{5.36}$$

This is the spontaneous emission power that would be displayed on an optical spectrum analyser with a wavelength resolution bandwidth $\Delta\lambda = c\Delta\nu/\nu^2$. The above analysis becomes much simpler for negligible facet reflectivities.

### 5.1.6 Simulations

The model can be used to determine how parameters such as signal gain and output noise depend on the photon and carrier density distributions in the amplifier. The exact distribution of these internal variables of course will depend on the amplifier input signals and operating conditions. To illustrate some of these dependencies, the steady-state performance of an SOA with parameters listed in Table 5.1 is modelled. In the numerical simulations the following settings are used: $N_z = 10$, $N_m = 40$ and $K_m = 20$.

Experimental and modelled SOA output spectra are shown in Fig. 5.3. The predicted gain ripple of 0.5 dB at the signal wavelength is identical to the experimental value. The predicted signal gain is 25.3 dB, which is close to the experimental value of 25.5 dB. The amplifier noise figure can be calculated using the formula [15]

$$NF = 10\log_{10}\left[\frac{\sigma_{ASE}}{h\nu G(\nu)} + \frac{\eta_{out}}{G(\nu)}\right] \quad (dB) \tag{5.37}$$

where $G(\nu)$ is the fibre-to-fibre gain at optical frequency $\nu$. $\sigma_{ASE}$ is given approximately by

$$\sigma_{ASE}(\nu) \approx \frac{(1-R_2)\eta_{out} h\nu N_j^+(L)}{K_m \Delta\nu_m} \tag{5.38}$$

where $N_j^+(L)$ is the closest forward propagating spontaneous emission photon rate at the output facet of the amplifier to the signal wavelength.

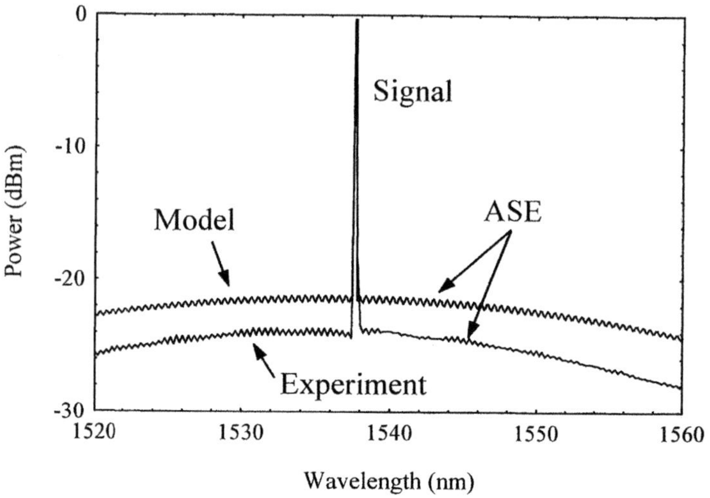


*Figure 5.3.* SOA output spectrum. The input signal has a wavelength of 1537.7 nm and power of -25.6 dBm. The bias current is 130 mA. The resolution bandwidth is 0.1 nm (after [8] with the permission of the IEEE).

A noise figure of $11.4 \pm 0.5$ dB at 1537.7 nm and a bias current of 130 mA is predicted compared to the experimental value of $8.8 \pm 0.3$ dB. This difference is primarily due to the length approximation of (5.1). Experimental and modelled gain versus bias current characteristics are shown in Fig. 5.4.

Predicted fibre-to-fibre gain and total output ASE power versus input signal power characteristics are shown in Fig. 5.5 The saturation output power in this case is 3.3 dBm. The form of these characteristics can be explained by reference to Fig. 5.6, which shows the carrier density, ASE and signal photon rate spatial distributions in the amplifier at low and high input signal powers. The backward propagating signal is negligible due to the low facet reflectivities. At low input powers, the carrier density has a symmetrical spatial distribution, peaking at the centre of the SOA and tailing off towards the input and output facets as shown in Fig. 5.6(a). This is because the ASE peaks in these regions as shown in Fig. 5.6(b). At high input powers the carrier density spatial distribution becomes more asymmetrical, as shown in Fig. 5.6(c), with the peak moving towards the input facet. This caused by the input signal dominating over ASE, as shown in Fig. 5.6(d).

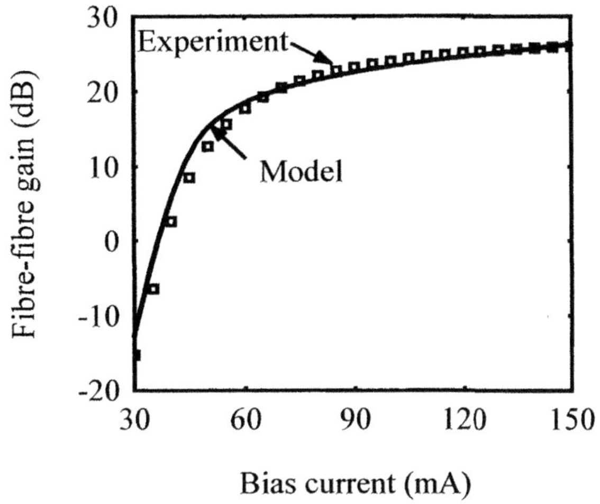


*Figure 5.4.* SOA experimental and modelled fibre-to-fibre gain versus bias current characteristic. The input signal power is -25.6 dBm and the signal wavelength 1537.7 nm (after [8] with the permission of the IEEE).

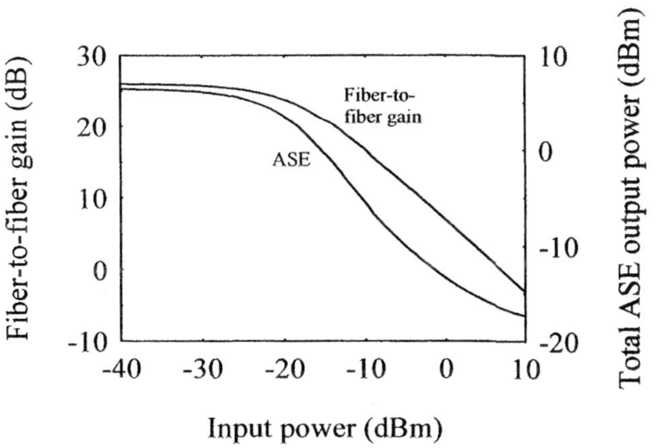


*Figure 5.5.* Simulated fibre-to-fibre gain and total output ASE power versus input signal power for a signal wavelength of 1537.7 nm and bias current of 130 mA (after [8] with the permission of the IEEE).

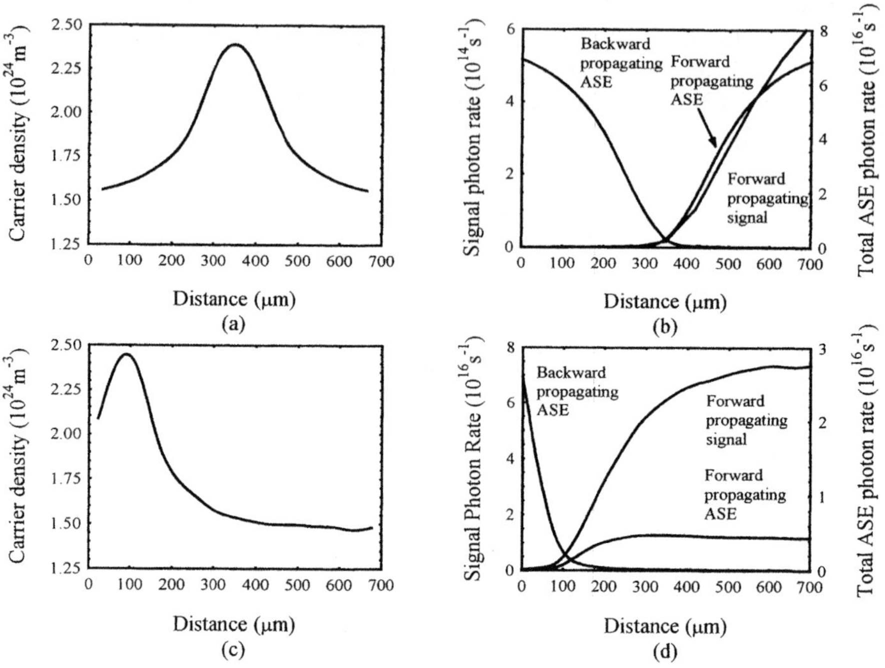


*Figure 5.6.* Spatial distributions of SOA carrier density, forward and backward propagating ASE and signal photon rates. Distance is measured from the SOA input facet. (a) and (b): -40 dBm input signal power. (c) and (d): -10 dBm input signal power. Saturation effects due to the amplified signal and ASE are evident. The signal wavelength is 1537.7 nm and the bias current is 130 mA (after [8] with the permission of the IEEE).

In Fig. 5.7 simulated SOA output ASE spectra are shown with the input signal power as parameter where the signal wavelength is near the unsaturated gain curve peak. At low signal powers, the carrier density is high leading to a high signal gain. As the signal power is increased, the carrier density decreases causing the material gain spectrum peak to shift to longer wavelengths so reducing the signal gain even further. This effect is called nonlinear gain compression. Fig. 5.7 also illustrates the fact that SOAs are homogeneously broadened devices, because variations in the signal power cause changes in the amplifier gain spectrum at all wavelengths.

## 5.2 DYNAMIC MODEL

As SOAs are usually used to amplify modulated light, it is of interest to model the dynamic performance of the amplifier. Dynamic SOA models can be used to predict pattern effects where amplified optical bits affect the following bits, intermodulation distortion where new frequency components are generated in an amplified analogue signal, and channel crosstalk that occurs when multiple signals are amplified. As is the case for SOA steady-state models, many dynamic SOA models with varying degrees of complexity and accuracy exist [16-20]. In this section three particular regimes are considered: (a) A basic dynamic model to analyse the performance of an SOA with digitally modulated input signals whose temporal properties are less than the transit time through the amplifier; (b) A model to predict the effects of amplification on ultrafast pulses and (c) a brief consideration of the effects of amplification on analogue signal transmission. In the following models it is assumed that the amplifier current is constant.

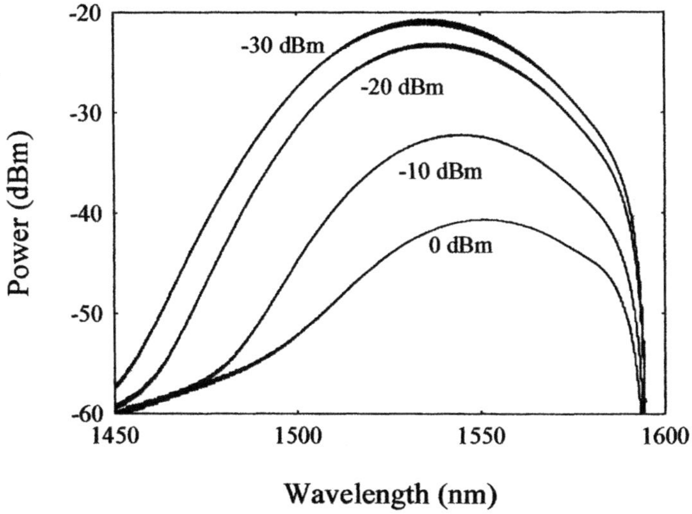


*Figure 5.7.* Simulated SOA output ASE spectra with the input signal power as parameter. The signal wavelength is 1537.7 nm and the bias current is 130 mA. The resolution bandwidth is 0.1 nm.

### 5.2.1 Basic dynamic model

In this section we consider a simple dynamic model for an SOA with digitally modulated inputs. It is assumed that the temporal properties of the input signals are on a time scale much greater than the amplifier transit time $\tau_{tr}$, given by

$$\tau_{tr} = \frac{Ln_1}{2c} \tag{5.39}$$

where $n_l$ is the active region refractive index and $L$ the amplifier length. Typical values of the transit time are of the order of picoseconds. Most conventional modulation schemes have time scales much greater than $\tau_r$.

We begin with a simplified version of the carrier density rate equation (5.29), ignoring spontaneous emission, approximating $g_m$ by $g_{m,peak}$ (equation (4.24)), transforming the signal photon rate to intensity and assuming a single input signal,

$$\frac{dn(z,t)}{dt} = \frac{I}{eV} - \frac{n(z,t)}{\tau_c} - \frac{a_1[n(z,t) - n_0]I_{sig}(z,t)}{hv} \tag{5.40}$$

where

$$I_{sig} = \frac{\Gamma h v N_{sig}}{A} \tag{5.41}$$

where $A$ is the active region cross-section area and $v$ the signal optical frequency. It is assumed that $\tau_c$ is constant. A simplified travelling-wave equation for the signal intensity is

$$\frac{dI_{sig}(z,t)}{dz} = \Gamma a_1[n(z,t) - n_0]I_{sig}(z,t) \tag{5.42}$$

(5.42) assumes that the light intensity responds instantaneously to temporal changes in the carrier density. The solution to (5.42) is

$$I_{sig}(z,t) = I_{sig,in}(t)\exp\left(\int_0^z \Gamma a_1[n(z',t) - n_0]dz'\right) \tag{5.43}$$

where $I_{sig,in}(t)$, the input signal intensity, is given by

$$I_{sig,in}(t) = \frac{\Gamma P_{in}(t)}{A} \tag{5.44}$$

where $P_{in}(t)$ is the input signal power. (5.40) and (5.43) can be solved numerically to determine the output signal intensity $I_{sig}(L,t)$ and the carrier density. A flow chart of the numerical algorithm is shown in Fig. 5.8. For a stable solution, the time step chosen should be an order of magnitude less than the carrier lifetime.

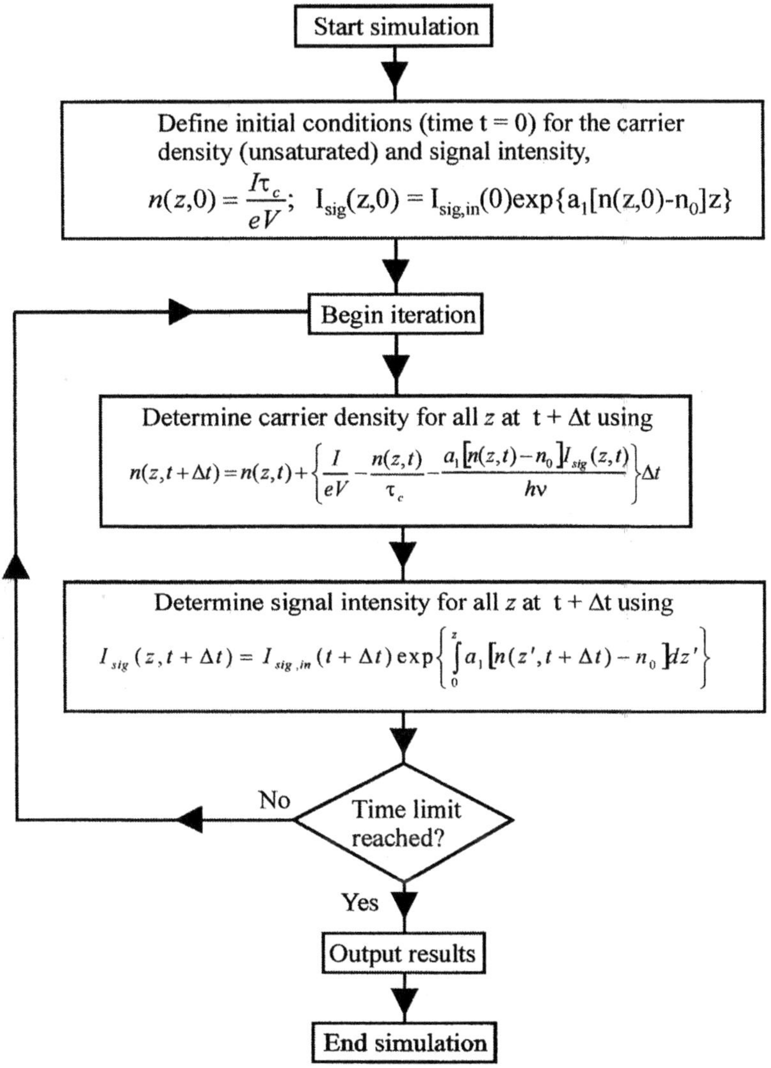


*Figure 5.8.* Flow chart of the SOA dynamic algorithm for temporal variations much larger than the transition time. The integral term can be evaluated numerically.

The effects of amplification by an SOA on an on-off keyed optical signal are shown in Fig. 5.9. When the amplifier is unsaturated the output signal suffers little distortion. This is because the carrier density is effectively independent of the input signal. In Fig. 5.9(a) the spatially averaged carrier density and output signal are shown for a signal repetition rate much less than $1/\tau_c$. A space is represented by zero intensity and a mark is represented by an intensity $I_{sig,1}$ comparable to the amplifier saturation intensity $I_{s,sat}$ (see equation (3.20)). In this case the rise and fall times of the carrier density

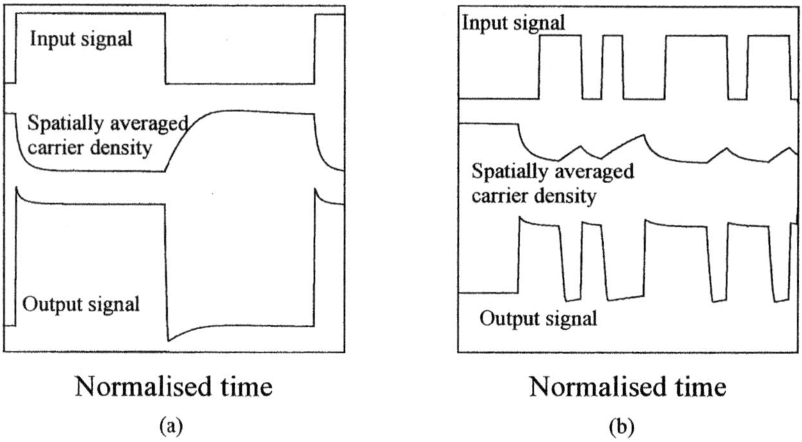


*Figure 5.9.* Dynamic simulation of waveform distortion and pattern effects in an SOA. (a) Signal repetition rate << $1/\tau_c$; (b) repetition rate comparable $1/\tau_c$.

response are much less than the input signal repetition rate. Although output signal distortion does occur, there are no pattern effects.

In Fig. 5.9(b) the signal repetition rate is of the same order as $1/\tau_c$ In this case the carrier density has not enough time to reach a steady-state value between signal transitions giving rise to pattern effects as well as signal distortion.

In the case where the repetition rate is much greater than $1/\tau_c$ and the bit pattern is random, there are negligible signal distortion or pattern effects. This is because the carrier density cannot keep pace with changes in the signal. It then settles at a level determined by the mean input signal power. The above algorithm can also be used to evaluate the crosstalk that occurs in multichannel amplification. Similar waveform distortion and pattern effects occur, the severity of which depends on the sum of the signal powers and on the individual signal repetition rates.

### 5.2.2 Fast pulse amplification

Because SOAs have large optical bandwidths (e.g. a wavelength bandwidth of 30 nm in the 1.55 μm region corresponds to a frequency bandwidth of 3.7 THz) they are able to amplify pulses as short as a few picoseconds. However, gain saturation in the amplifier leads to pulse distortion. When an SOA is used to amplify optical pulses with temporal characteristics of the order of the amplifier transit time (i.e. picoseconds), the model described in Section 5.2.1 is not adequate. In this case it is necessary to model pulse propagation through the amplifier with a travelling wave equation containing spatial and time derivatives. In the following model we follow the method of [21], where the main assumption is that the pulse width is greater than the intraband relaxation time (usually < 0.1 ps). Assuming that the input light signal is linearly polarised and remains so during propagation, the longitudinally varying component of the signal electric field $E(z,t)$ obeys the partial differential equation

$$\frac{\partial E}{\partial z} + \frac{1}{v_g}\frac{\partial E}{\partial t} = \frac{(1 - j\alpha_l)gE}{2} \tag{5.45}$$

The signal field can be expressed as

$$E(z,t) = \sqrt{P(z,t)}e^{j\phi(z,t)} \tag{5.46}$$

where $P$ and $\phi$ are the signal power and phase respectively. $v_g$ is the group velocity and $\alpha_l$ is a constant usually referred to as the linewidth enhancement factor. It couples the phase of the propagating signal to the gain coefficient. To simplify the solution of (5.45) the *static* time variable $t$ and spatial coordinate $z$ can be transformed into a time frame $\tau$ *moving* with the propagating signal, i.e.

$$\tau = t - \frac{z}{v_g} \tag{5.47}$$

Using this transformation and inserting (5.46) into (5.45) gives

$$\begin{aligned}\frac{\partial P}{\partial z} &= gP \\ \frac{\partial \phi}{\partial z} &= -\frac{1}{2}\alpha_l g\end{aligned} \tag{5.48}$$

The carrier density rate equation used is

$$\frac{\partial n}{\partial t} = \frac{I}{edWL} - \frac{n}{\tau_c} - \frac{gP}{dWh\nu} \tag{5.49}$$

The gain coefficient $g$ is approximated by

$$g = \Gamma a_1 (n - n_0) \tag{5.50}$$

(5.49) can then be transformed into a rate equation for $g$,

$$\frac{\partial g}{\partial \tau} = \frac{(g - g_0)}{\tau_c} - \frac{gP}{E_{sat}} \tag{5.51}$$

where $g_0$ is the small signal gain coefficient given by

$$g_0 = \Gamma a_1 n_0 \left( \frac{I}{I_0} - 1 \right) \tag{5.52}$$

$I_0$, the bias current required for transparency, is given by

$$I_0 = \frac{eVn_0}{\tau_c} \tag{5.53}$$

The saturation energy is

$$E_{sat} = \frac{h\nu dW}{\Gamma a_1} \tag{5.54}$$

The solution of (5.51) is

$$P_{out}(\tau) = P_{in}(\tau)\exp[h(\tau)] \tag{5.55}$$

with

$$h(\tau) = \int_0^\tau g(z,\tau)dz \tag{5.56}$$

$h(\tau)$ is the integrated gain at each point of the propagating pulse profile. $P_{out}(\tau)$ and $P_{in}(\tau)$ are the input and output pulse powers respectively. Integrating (5.51) and using (5.56) gives an ordinary differential equation for $h$,

$$\frac{dh}{d\tau} = \frac{g_0 L - h}{\tau_c} - \frac{P_{in}(\tau)}{E_{sat}}\left(e^h - 1\right) \tag{5.57}$$

For a given input pulse shape and gain $g_0L$, (5.57) can be solved to obtain $h(\tau)$. $P_{out}$ can then be found using (5.55).

(5.57) can be solved analytically in one important case. If the pulse width $\tau_p << \tau_c$ the first term on the RHS of (5.57) can be neglected. This is because the pulse width is so short that the gain has no time to recover. In this case the solution of (5.57) is

$$h(\tau) = -\ln\left\{1 - \left(1 - \frac{1}{G_0}\right)\exp\left[-\frac{U_{in}(\tau)}{E_{sat}}\right]\right\} \tag{5.58}$$

where $G_0 = \exp(g_0 L)$ is the unsaturated single-pass amplifier gain and

$$U_{in}(\tau) = \int_{-\infty}^{\tau} P_{in}(\tau')d\tau' \tag{5.59}$$

$U_{in}$ is the energy contained in the pulse up to time $\tau$. $U_{in}(\infty)$ is the total pulse energy $E_{in}$. As an example consider a Gaussian input pulse shape

$$P_{in}(\tau) = \frac{E_{in}}{\tau_0\sqrt{\pi}}\exp\left(-\frac{\tau^2}{\tau_0^2}\right) \tag{5.60}$$

$\tau_0$ is related to the pulse full width at half maximum $\tau_p$ by $\tau_p \approx 1.665\tau_0$. Inserting (5.60) in (5.59) gives

$$U_{in}(\tau) = \frac{1}{2} E_{in}\left[1 + erf\left(\tau/\tau_0\right)\right] \tag{5.61}$$

where *erf* (Chapter 6) is the error function. The instantaneous amplifier gain is

$$G(t) = \exp[h(\tau)] = \left\{1 - \left(1 - \frac{1}{G_0}\right)\exp\left[-\frac{U_{in}(\tau)}{E_{sat}}\right]\right\}^{-1} \tag{5.62}$$

The leading edge of the pulse experiences a gain $G(\infty) = G_0$ as the amplifier is not yet saturated. The trailing edge of the pulse experiences a gain $G_f$ given by

$$G_f = G(-\infty) = \frac{G_0}{G_0 - (G_0 - 1)\exp\left(-E_{in}/E_{sat}\right)} \tag{5.63}$$

The quantity of most practical interest is the energy gain given by

$$G_E \equiv \frac{E_{out}}{E_{in}} = \frac{1}{E_{in}} \int_{-\infty}^{\infty} P_{in}(\tau)\exp[h(\tau)]d\tau \tag{5.64}$$

Using (5.62) $G_E$ can be expressed in terms of $G_0$ and $G_f$ as

$$G_E = \frac{\ln\left(\dfrac{G_0 - 1}{G_f - 1}\right)}{\ln\left(\dfrac{G_0 - 1}{G_f - 1}\right) - \ln\left(\dfrac{G_0}{G_f}\right)} \tag{5.65}$$

(5.65) shows that the energy gain is independent of the input pulse shape. For a CW input signal the amplifier gain $G_{cw}$ can be got from (3.24),

$$G_{cw} = G_0 \exp\left[-\frac{(G_{CW} - 1)P_{in}}{P_{sat}}\right] \tag{5.66}$$

where $P_{sat} = E_{sat}/\tau_c$ .

The saturated CW and energy gains are plotted as a function of $E_{in}/E_{sat}$ with the unsaturated gain as parameter in Fig. 5.10(a), showing that the energy gain saturates faster than the CW gain. In Fig. 5.10(b) typical output pulse shapes are shown with the unsaturated gain as parameter. The output pulse shape become more asymmetric as $G_0$ increases with the leading edge becoming sharper than the tailing edge due to the dynamic gain difference. The pulse spectrum is also affected by gain saturation. These effects are primarily due to the phenomenon of self phase modulation, discussed in Chapter 7.

### 5.2.3 Analog signal dynamics

SOAs can also be used as amplifiers in analog communication systems such as cable TV distribution networks. However, analog system performance is very sensitive to signal distortion. This requires that amplifiers used in such systems must be highly linear. In particular gain saturation can cause both harmonic and phase distortion of analogue modulated signals. Significant distortion can occur at input signal powers well below the amplifier saturation input power. Conventional bulk active region SOAs do not meet the linearity requirements of analog systems [22]. However, it has been shown [23] that gain-clamped MQW SOAs, due to their superior gain saturation characteristics, can meet the stringent requirements of analog systems.

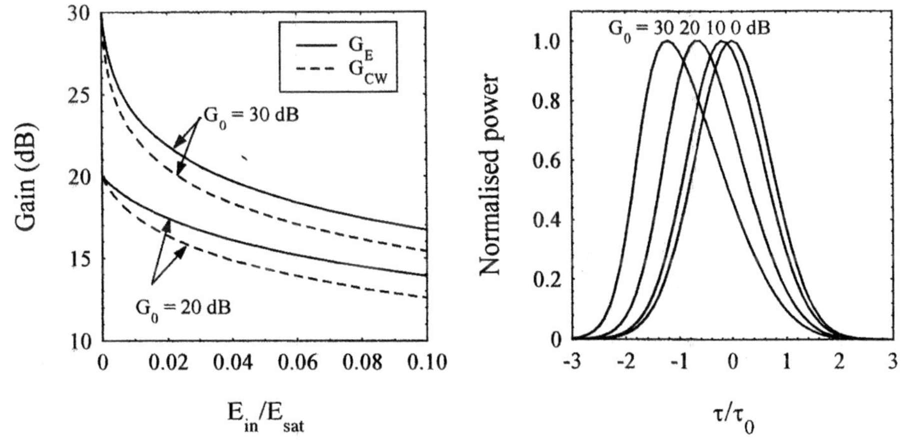


*Figure 5.10.* (a) SOA saturated CW and energy gains. The x-axis corresponds to $P_{in}/P_{sat}$ in the case of $G_{cw}$. (b) Output pulse shapes with $G_0$ (dB) as parameter for a Gaussian input pulse shape (after [21] with the permission of the *IEEE*).

## REFERENCES

1. L. Gillner, Comparative study of some travelling-wave semiconductor laser amplifier models, *IEE Proc. Pt. J. Optoelectronics*, **139**, 339-347 (1992).
2. P. Brosson, Analytical model of a semiconductor optical amplifier, Brosson, *J. Lightwave Technol.*, **12**, 49-54 (1994).
3. D. Marcuse, Computer model of an injection laser amplifier, *IEEE J. Quantum Electron.*, **19**, 63-73 (1983).
4. A.J. Lowery, Requirements for models of semiconductor optical amplifiers, *IEE Colloquium on Optical Amplifiers for Communications*, 18/1-18/4 (1989).
5. I.D. Henning, M.J. Adams and J.V. Collins, Performance predictions from a new optical amplifier model, *IEEE J. Quantum Electron.*, **21**, 609-613 (1985).
6. A.J. Lowery, A qualitative comparison between two semiconductor laser amplifier equivalent circuit models, *IEEE J. Quantum Electron.*, **26**, 1369-1357 (1990).
7. L. Gillner, E. Goobar, L. Thylen and M. Gustavsson, Semiconductor laser amplifier optimization: an analytical and experimental study, *IEEE J. Quantum Electron.*, **25**, 1822 - 1827 (1989).
8. M.J. Connelly, Wideband SOA steady-state numerical model, *IEEE J. Quantum Electron.*, **37**, 439-447 (2001).
9. C. Deguet, D. Delprat, G. Crouzel, N.J. Traynor, P. Maigne, T. Pearsal, C. Lerminiaux, N. Andreakis, C. Caneau, F. Favire, R. Bhat and C.E. Zah, Homogeneous Buried Ridge Stripe Semiconductor Optical Amplifier with near Polarisation Independence, *Proc. European Conference on Optical Communications* (1999).
10. S. Adachi, *Physical properties of III-IV compounds*, John Wiley and Sons, New York: Chichester (1992).
11. M.J. Connelly and R.F. O'Dowd, Travelling-wave Semiconductor Optical Amplifier Detector Noise Characteristics, *IEE Proc. Pt. J. Optoelectronics.*, **142**, 23-28 (1995).
12. A. Yariv, *Optical electronics*, HRW International (1985).
13. M-C Amann and W. Thulke, Current confinement and leakage currents in planar buried-ridge-structure laser diodes on n-substrate," *IEEE J. Quantum Electron.*, **25**, 1595-1602 (1989).
14. W. H. Press, B. P. Flannery, S. A. Teukolsky and W. T. Vetterling, *Numerical Recipes in Pascal*, Cambridge University Press, Cambridge (1994).
15. J. C. Simon, P. Doussiere, L. Pophillat and B. Fernier, Gain and noise characteristics of a 1.5μm near-travelling-wave semiconductor laser amplifier, *Electron. Lett.*, **25**, 434-436 (1989).
16. A. J. Lowery, New inline wideband dynamic semiconductor laser amplifier model, *IEE Proc. Pt. J. Optoelectronics.*, **135**, 242-250 (1988).
17. T. Durhuus, B. Mikkelsen, K. E. Stubkjaer, Detailed Dynamic model for semiconductor optical amplifiers and their crosstalk and intermodulation distortion, *J. Lightwave Technol.*, **10**, 1056-1064 (1992).
18. J. Chen, D. Li, H. Lu., Dynamic amplification of coded pulses by a travelling-wave semiconductor laser amplifier, *J. Opt. Comm.*, **9**, 133-139 (1988).
19. H. Tomofuji, N. Nishimoto, T. Horimatsu, T. Minami and T. Touge, Cumulative waveform distortion in cascaded optical amplifier repeaters for multigigabit IM/DD systems, *IEEE Photon. Technol. Lett.*, **2**, 756-758 (1990).
20. K. Inoue, Waveform distortion in a gain saturated semiconductor optical amplifier for NRZ & RZ formats, Inoue, *IEE Proc. Pt. J. Optoelectronics*, **144**, 433-437 (1997).

21. G. P. Agrawal and N. A. Olsson, Self phase modulation and spectral broadening of optical pulses in semiconductor laser amplifiers, *IEEE J. Quantum Electron.*, **25**, 2297-2306 (1989).
22. J. A. Constable, I. H. White, A. N. Coles and D. G. Cunningham, Harmonic and phase distortion of analogue amplitude-modulated signals in bulk near travelling-wave semiconductor optical amplifiers, *IEE Proc. Pt. J. Optoelectronic.*, **139**, 389-398, (1992).
23. L. F. Tiemeijer, P. J. Thijs, T. v. Dongen, J. J. M. Binsma, E. J. Jansen and H. R. J. R. van Helleputte, Reduced intermodulation distortion in 1300nm gain-clamped MQW laser amplifiers, *IEEE Photon. Technol. Lett.*, **7**, 284-286 (1995).

# Chapter 6

# BASIC NETWORK APPLICATIONS

The principal applications of SOAs in optical communication systems can be classified into three main areas: Postamplifier or booster amplifier to increase transmitter laser power, in-line amplifier to compensate for fibre and other transmission losses in medium and long-haul links and preamplifier to improve receiver sensitivity. The incorporation of optical amplifiers into optical communication links can improve system performance and reduce costs. SOAs can also be used to perform other functions such as high speed optical switching, wavelength conversion and in-line detection. Such functional applications are discussed in Chapter 7.

In this chapter we focus on conventional applications of SOAs in optical communication links and networks. We begin with a brief review of the essentials of direct and coherent optical detection followed by a derivation of the detection statistics of amplified light. Next the application of SOAs as post-amplifiers, preamplifiers and in-line amplifiers are considered. The properties of cascaded in-line SOAs are analysed. The performance of SOAs in WDM networks is discussed along with techniques that can be utilised to reduce crosstalk and intermodulation distortion. Finally some remarks are made concerning the use of SOAs in analog transmission systems.

## 6.1 REVIEW OF DIRECT AND COHERENT OPTICAL COMMUNICATIONS

Optical communication systems can be classified broadly as non-coherent and coherent transmission [1-3]. In non-coherent transmission only the intensity of an optical carrier signal is modulated. At the receiver the signal is directly detected, a process that is only sensitive to the signal intensity. Such systems are termed Intensity Modulation-Direct Detection (IM-DD).

In coherent systems a local oscillator light signal is added to the incoming optical carrier signal prior to detection. In such schemes the phase or frequency of the optical carrier, as well as its intensity, can be detected. In optical communications the term coherent is taken to mean any system where a local oscillator laser is added to the incoming modulated light signal prior to detection, even if the phase or frequency of the incoming light signal is ignored in subsequent processing and demodulation.

### 6.1.1 Intensity modulation - direct detection receiver

A schematic diagram of a basic IM-DD receiver is shown in Fig. 6.1(a). In this scheme an intensity modulated optical carrier signal is detected by a photodetector (p-i-n diode or avalanche photodiode (APD)). The resulting photocurrent is amplified and passed to a decision circuit that determines whether each received bit is a mark or space.

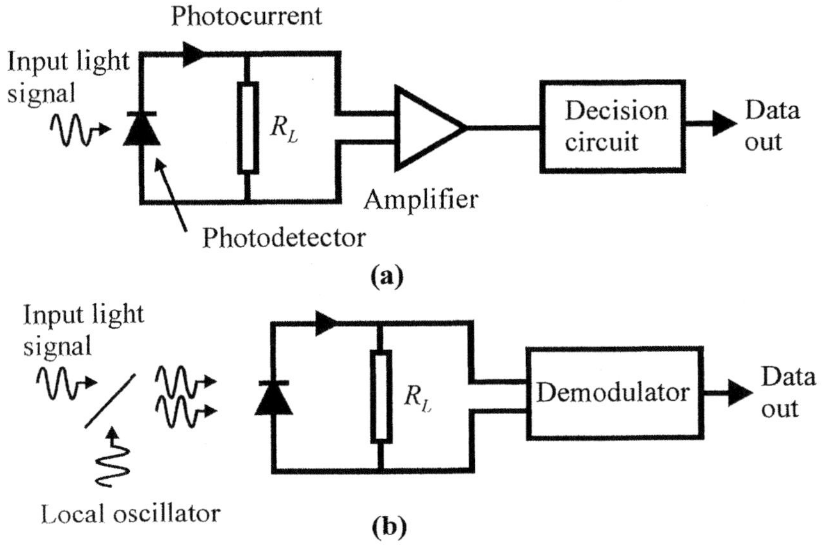


*Figure 6.1* Block diagrams of basic (a) IM-DD and (b) coherent optical receivers.

Two methods can be used to make decisions on the received waveform. The first approach involves sampling the received waveform every bit period, usually at the centre of the bit and comparing the sampled value to a threshold level. If the sampled value is less than the threshold level the received bit is interpreted as a space and vice versa. The second approach is the 'integrate and dump' technique where the received waveform is integrated over a bit period. The output of the integrator is then compared to a threshold level and an identical decision made as in the first method. The former approach has the advantage of minimising the effects of intersymbol interference (ISI) while the latter, because of the integration operation, tends to minimise the effects of noise. Both reception techniques require the generation of a local clock for synchronisation purposes.

The usual figure of merit for a digital receiver is the bit-error-rate (BER). The IM-DD receiver can be analysed as follows. We assume on-off keyed (OOK) modulation where spaces and marks are represented by input powers of zero and $2P_s$ respectively. $P_s$ is the average received power assuming that the transmission probabilities of a mark or a space are equal. The input signal field to the photodetector can be expressed as

$$E(t) = \begin{cases} E_1(t) = \sqrt{2P_s}Cos(2\pi v\,t) & \textit{for a mark} \\ E_0(t) = 0 & \textit{for a space} \end{cases} \tag{6.1}$$

where $v$ is the optical carrier frequency and $t$ time. The photocurrent $i_d$ from a photodetector is proportional to the square of the incoming light signal field averaged over an optical period, i.e.

$$i_d = R\left\langle E(t)^2 \right\rangle \tag{6.2}$$

where the < > operator indicates time averaging over an optical period and $R$ is the photodetector responsivity (A/W) given by

$$R = \frac{\eta e}{h v} \tag{6.3}$$

where $\eta$ is the detector quantum efficiency. Inserting (6.1) into (6.2) gives an equivalent signal photocurrent,

$$I_s(t) = \begin{cases} I_{s1} = 2RP_s & \textit{for a mark} \\ I_{s0} = 0 & \textit{for a space} \end{cases} \tag{6.4}$$

In (6.4) the term at twice $v$ has been neglected as it greatly exceeds the detector electrical bandwidth. In addition to the signal photocurrent there are also noise currents due to the detector dark current $I_d$ (a small photocurrent that exists even if there is no detected light) with variance $\overline{i}_d^2$, the signal shot noise current with variance $\overline{i}_s^2$ (due to the quantum nature of the detected light signal) and receiver circuit noise with variance $\overline{i}_c^2$. The latter noise source is mainly due to receiver thermal noise. These noise current variances can be expressed as

$$\begin{aligned} \overline{i}_d^2 &= 2eB_e I_d \\ \overline{i}_s^2 &= 2eB_e I_s \\ \overline{i}_c^2 &= \frac{4kTFB_e}{R_L} \end{aligned} \tag{6.5}$$

where $B_e$ is the receiver post-detection bandwidth, $F$ the receiver noise figure, $R_L$ the load resistance and $T$ the receiver temperature. For OOK modulation the total noise current variances for a mark and space are then given by

$$\sigma_{iT}^2 = \begin{cases} \sigma_{i1}^2 = \overline{i}_d^2 + \overline{i}_c^2 + 2eB_e I_{s1} & \text{mark} \\ \sigma_{i0}^2 = \overline{i}_d^2 + \overline{i}_c^2 & \text{space} \end{cases} \tag{6.6}$$

For a conventional OOK receiver (using the first approach above), if it is assumed that the noise currents have Gaussian probability density functions, the BER is given by [2]

$$BER = \frac{1}{2} erfc\left(\frac{Q}{\sqrt{2}}\right) \tag{6.7}$$

where the $Q$-factor is given by

$$Q = \frac{\sqrt{S_1} - \sqrt{S_0}}{\sqrt{\sigma_{i1}^2} + \sqrt{\sigma_{i0}^2}} \tag{6.8}$$

$S_1 = I_{s1}^2$ and $S_0 = 0$ are the mark and space signal powers respectively. *erfc* is the complementary error function given by

$$erfc(x) = 1 - erf(x) = \frac{2}{\sqrt{\pi}} \int_x^\infty e^{-u^2} du \tag{6.9}$$

where *erf* is the error function. *erfc* frequently occurs in the BER expressions of many types of receiver. Plots of *erfc* and BER versus $Q$ are shown in Fig. 6.2. For large values of $x$ the following approximation is useful,

$$erfc(x) \approx \frac{e^{-x^2}}{x\sqrt{\pi}}, \quad x \text{ large} \tag{6.10}$$

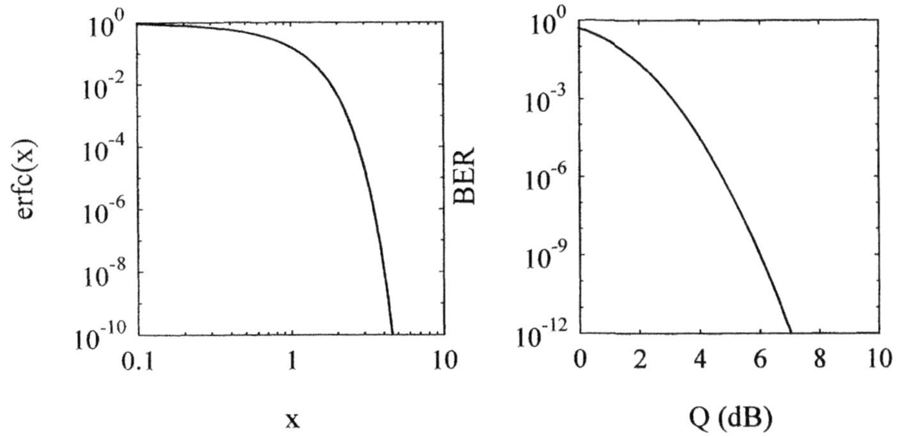


*Figure 6.2.* The complementary error function *erfc* and BER versus the *Q*-factor for an IM-DD receiver.

The receiver sensitivity can be defined as the average received optical power or photons per bit required for a given BER. Optical communication systems normally specify that the BER be $< 10^{-9}$, which requires that $Q > 6$. In the ideal case where dark current and circuit noise are negligible, then

$$Q^2 = \frac{S_1}{\sigma_{i1}^2} = \frac{RP_s}{eB_e} \tag{6.11}$$

If the photodetector has unity quantum efficiency and electrical bandwidth $B_e = 1/B_T$ where $B_T$ is the bit rate, then

$$Q^2 = 2N_b \tag{6.12}$$

where $N_b = P_s/(h\nu B_T)$ is the average number of photons per bit. This value is referred to as the quantum limit (or shot-noise limit). A BER of $10^{-9}$ requires that $N_b > 18$. If the received signal wavelength and bit rate are known, this number can be used to obtain the required minimum receiver average optical power or sensitivity. However in practical receivers it is almost impossible to approach the quantum limit. This is caused by circuit noise, which is normally greater than shot noise by many orders of magnitude, giving sensitivities in the range of hundreds to thousands of photons per bit. The above analysis has also neglected the effect of optical fibre dispersion. Fibre dispersion leads to pulse broadening and hence to ISI. This increases the BER and thereby reduces receiver sensitivity. As we shall see in Section 6.3.2 the sensitivity of practical IM-DD optical receivers can be greatly improved using optical preamplifiers. More detailed analyses of optical receiver sensitivity and BER can be found in [1-4].

### 6.1.2 Coherent detection

A schematic diagram of a basic coherent receiver is shown in Fig. 6.1(b). In this receiver type a local oscillator laser is added to the incoming modulated optical carrier signal. Both signals are then detected simultaneously by a photodetector. The resulting photocurrent is demodulated and processed to retrieve the data signal. There are two main advantages that coherent optical communications has compared to conventional IM-DD systems. Firstly, greater receiver sensitivities are attainable (as long as the signal and local oscillator phase noise are kept within acceptable limits) and secondly much greater receiver selectivity is possible. The first advantage implies that it is possible to increase the distance between regenerative repeaters in coherent systems compared to IM-DD systems. However, this advantage has been largely negated through the use of optical in-line amplifiers and preamplifiers in IM-DD systems. This has meant that interest in the commercial implementation of coherent optical transmission systems has faded, compared to the great flurry of activity in the 1980s before the advent of high performance and reliable optical fibre amplifiers. The second advantage means that it is possible using optical frequency division multiplexing (OFDM) to transmit many more optical channels, with very fine frequency spacing, on a single optical fibre. This leads to more efficient utilisation of the fibre bandwidth. The channel selection is carried out post-detection in the electrical domain. In IM-DD systems wavelength division multiplexing (WDM) can also be used to increase utilisation of the fibre bandwidth; however in this case channel selection is carried out prior to detection using optical filters or wavelength demultiplexers. The channel spacing is much greater compared to FDM. In any case the use of IM-DD systems greatly exceeds that of coherent systems. For this reason in the remainder of the book we concentrate on IM-DD systems.

## 6.2 AMPLIFIED LIGHT DETECTION STATISTICS

The addition of spontaneous emission (i.e. noise) is an inevitable consequence of the amplification of light. The use of an optical filter at the amplifier output can greatly reduce this noise, however it is impossible to eliminate it entirely. When the signal and accompanying noise are detected by a photodetector the square-law detection process gives rise to beat-noise currents in addition to the usual shot-noise. In the following analysis, based on the method of [5], expressions are derived for the noise components of the photocurrent generated by an optically amplified signal. The scenario under consideration is shown in Fig. 6.3, where a linearly polarised signal and accompanying ASE are passed through a polariser and an optical filter prior to detection.

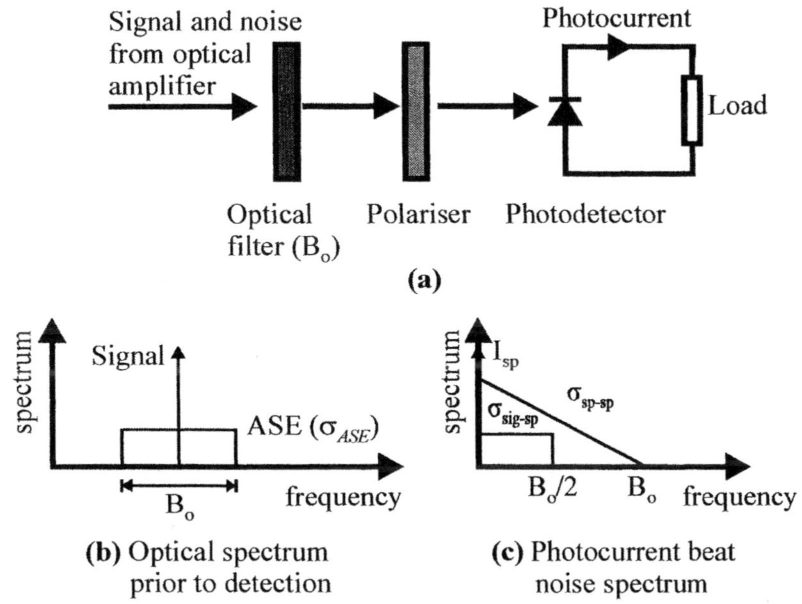


*Figure 6.3.* (a) Detection of amplified light and ASE. (b) Detector input optical spectrum. (c) Photocurrent beat noise spectrum.

The optical filter has a rectangular transfer function with bandwidth $B_o$ (Hz) centred on the signal wavelength. The photodetector has responsivity $R$ and post-detection bandwidth $B_e$ The function of the polariser is to pass the linearly polarised signal and reduce by half the ASE noise. This is because each spontaneously emitted photon can exist in one of two mutually orthogonal polarisations (assuming polarisation insensitive optical amplification). The ASE noise at the detector is assumed to have a uniform optical power spectral density $\sigma_{ASE}$, at a single polarisation, over the filter bandwidth. This assumption is valid as the optical filters used in practice have bandwidths much narrower than typical SOA spectral bandwidths. The signal component of the electric field at the detector can be expressed as

$$E_{sig}(t) = \sqrt{2P_{sig}}\,Cos(\omega_{sig}t) \tag{6.13}$$

where $P_{sig}$ and $\omega_{sig}$ are the signal power and optical angular frequency respectively. The ASE component of the input electric field can be expressed as

$$E_{sp}(t) = \sum_{i=-B_o/2\delta v}^{B_o/2\delta v} \sqrt{2\sigma_{ASE}\delta v}\,Cos\left[(\omega_{sig} + 2\pi i\delta v)t + \phi_i\right] \tag{6.14}$$

where $\delta v$ is a small frequency interval and $\phi_i$ is a random phase. The total detected electric field is then

$$E(t) = E_{sig}(t) + E_{sp}(t) \tag{6.15}$$

The resulting photocurrent using (6.2) is then

$$\begin{aligned} i_d &= R\left\langle\left[E_{sig}(t) + E_{sp}(t)\right]^2\right\rangle \\ &= R\left[\left\langle E_{sig}^2(t)\right\rangle + 2\left\langle E_{sig}(t)E_{sp}(t)\right\rangle + \left\langle E_{sp}^2(t)\right\rangle\right] \end{aligned} \tag{6.16}$$

The first term on the RHS of (6.16) is simply the detected signal photocurrent $I_s (= RP_{in})$. It is a shot-noise process with variance

$$\bar{i}_{sig-shot}^2 = 2eB_eRP_{sig} \tag{6.17}$$

The second term on the RHS of (6.16) results from the beating of the signal with the ASE. It can be written as

$$i_{sig-sp} = 2R\sqrt{P_{in}\sigma_{sp}\delta v}\sum_{i=-M}^{M} Cos(2\pi i\delta v\, t + \phi_i) \tag{6.18}$$

where $M = B_o/(2\delta v)$ and the terms about frequency $2\omega_{sig}$ have been neglected. Each frequency $2\pi i\delta v$ in (6.18) has two components with random phase. This implies that the power spectrum of $i_{sig\text{-}sp}$ is uniform in the frequency interval $[0,B_o/2]$ with power spectral density ($A^2/Hz$)

$$\sigma^2_{sig-sp} = 4R^2 P_{sig}\sigma_{ASE} \tag{6.19}$$

Hence the signal-spontaneous beat noise current variance is

$$\bar{i}^2_{sig-sp} = 4R^2 P_{sig}\sigma_{ASE} B_e \tag{6.20}$$

The third term on the RHS of (6.16) results from the beating of the spontaneous noise with itself and is given by

$$i_{sp-sp} = 2R\sigma_{ASE}\delta v \left\langle \sum_{i=-M}^{M} \cos(\beta_i) \sum_{j=-M}^{M} \cos(\beta_j) \right\rangle \tag{6.21}$$

with

$$\begin{aligned} \beta_i &= (\omega_{sig} + 2\pi i\delta v)t + \phi_i \\ \beta_j &= (\omega_{sig} + 2\pi j\delta v)t + \phi_j \end{aligned} \tag{6.22}$$

Evaluating (6.21), ignoring terms with frequencies about $2\omega_{sig}$, gives

$$i_{sp-sp} = R\sigma_{ASE}\sum_{i=0}^{2M}\sum_{j=0}^{2M} \cos\left[(i-j)2\pi\delta v\, t + \phi_i - \phi_j\right] \tag{6.23}$$

The term at zero frequency is obtained for $i = j$. There are $2M$ such terms giving a mean spontaneous current

$$I_{sp} = R\sigma_{ASE} B_o \tag{6.24}$$

This current is a shot noise process with variance

$$\overline{i}^2_{sp-shot} = 2eB_e I_{sp} \tag{6.25}$$

The remaining terms in (6.23) can be organised according to their frequencies. The number of terms at frequency $l\delta v$ is $2M-l$ where $l$ is an integer ranging between $-(2M-1)$ and $-1$ and between $1$ and $2M+1$. Terms with the same absolute frequency add in phase so the single-sided power spectrum density $\sigma^2_{sp-sp}$ of $i_{sp-sp}$ extends from 0 to $B_o$ with a triangular shape. As $B_o$ is much larger than $B_e$ we need only be concerned with the power spectral density near zero frequency so

$$\sigma^2_{sp-sp} \approx 2R^2 \sigma^2_{ASE} B_o \tag{6.26}$$

Hence the spontaneous-spontaneous beat noise current variance is

$$\overline{i}^2_{sp-sp} = 2R^2 \sigma^2_{ASE} B_o B_e \tag{6.27}$$

The total photocurrent variance is the sum of the above current variances, i.e.

$$\overline{i}^2_d = \overline{i}^2_{sig-shot} + \overline{i}^2_{sp-shot} + \overline{i}^2_{sig-sp} + \overline{i}^2_{sp-sp} \tag{6.28}$$

In practical receivers the polarisation of the input signal is usually not known so a polariser cannot be used. In this case the spontaneous shot noise and spontaneous-spontaneous beat current variances must be doubled. The signal-spontaneous beat noise current variance is not affected as the signal can only beat with noise having the same polarisation as itself.

### 6.2.1 Optical amplifier electrically equivalent noise figure

A useful figure of merit for an optical amplifier is the electrically equivalent noise figure $F$, defined as the ratio between the amplifier input and output electrical SNRs, i.e.

$$F = \frac{(S/N)_{in}}{(S/N)_{out}} \tag{6.29}$$

The SNRs are calculated by assuming that the amplifier input signal and output signal plus ASE are passed through a narrowband optical filter prior to detection by an ideal photodetector (i.e. unity quantum efficiency). In this case the only photocurrent noise terms that need to be taken into account are the signal shot noise and the signal-spontaneous beat noise. If the input signal power to the amplifier is $P_{sig}$, then the electrically equivalent input SNR is

$$\left(\frac{S}{N}\right)_{in} = \frac{(RP_{sig})^2}{2eB_e RP_{sig}} \tag{6.30}$$

The amplifier output electrically equivalent SNR is

$$\left(\frac{S}{N}\right)_{out} = \frac{(GRP_{sig})^2}{2eB_e GRP_{sig} + 4R^2 GP_{sig}\sigma_{ASE}B_e} \tag{6.31}$$

where $G$ is the amplifier gain. Taking the ratio of (6.30) and (6.31) gives

$$F = \frac{1}{G} + \frac{2\sigma_{ASE}}{h\nu G} \tag{6.32}$$

where $\nu$ is the signal frequency. From (2.24) $\sigma_{ASE}$ can be expressed as

$$\sigma_{ASE} = n_{sp}h\nu(G-1) \tag{6.33}$$

where $n_{sp}$ is the spontaneous emission factor. If $G >> 1$ the noise figure becomes

$$F = 2n_{sp} \tag{6.34}$$

The minimum value possible for $n_{sp}$ is 1, so the noise figure of an ideal optical amplifier is 2 (i.e. 3 dB). If the amplifier has an input coupling efficiency $\eta_{in}$, the noise figure is

$$F = \eta_{in}\left(\frac{1}{G} + \frac{2\sigma_{ASE}}{h\nu G}\right) \tag{6.35}$$

The noise figure is not degraded by the amplifier output coupling losses. If a polariser is not used, the minimum noise figure is 6 dB and the second term on the RHS of (6.35) must be doubled.

## 6.3 SOA GAIN BLOCKS IN OPTICAL NETWORKS

Now that we have analysed the properties of detected amplified light, we are in a position to investigate the application of SOAs as basic gain blocks in optical communication systems. These applications, shown in Fig. 6.4 are the post-amplifier or booster amplifier, in-line amplifier and preamplifier. The main requirements of optical amplifiers for such applications are listed in Table 6.1.

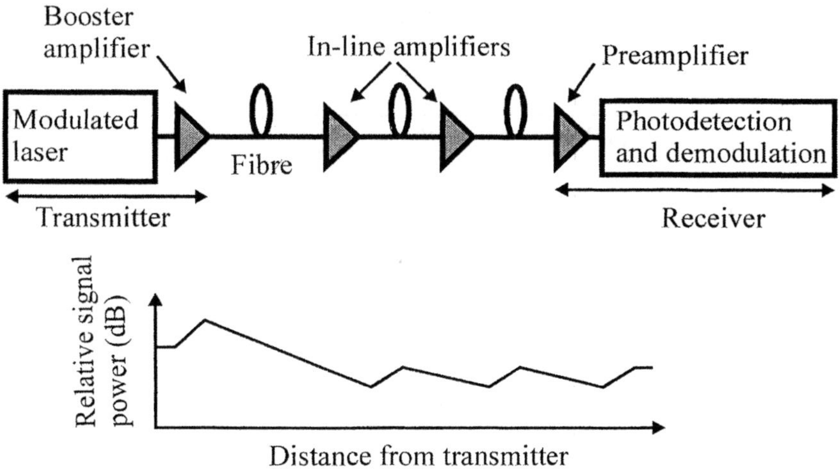


*Figure 6.4.* Application of SOAs as booster amplifier, in-line amplifiers and preamplifier in an optical transmission link.

*Table 6.1.* Optical amplifier requirements in optical transmission systems.

| | Post-amplifier | In-line amplifier | |
|---|---|---|---|
| High gain | Yes | Yes | Yes |
| High saturation output power | Yes | Yes | Not critical |
| Low noise figure | Not critical | Yes | Yes |
| Low polarisation sensitivity | Not critical | Yes | Yes |
| Low insertion loss | Not critical | Yes | Yes |
| Optical filter | Not necessary | Not critical | Yes |
| Optical isolators | Yes | Not critical | Not critical |

### 6.3.1 Booster amplifier

The function of a booster amplifier is to increase a relatively high power input signal prior to transmission. The main applications of booster amplifiers are listed in Table 6.2. Boosting laser power in an optical transmitter enables the construction of medium-haul links with increased transmission distance. Such links simply consist of an optical fibre between the transmitter and receiver. As this involves no active components in the transmission link, reliability and performance are improved.

In long-haul links the use of a booster amplifier can increase the link power budget and perhaps reduce the number of in-line amplifiers or regenerators required. Booster amplifiers are also useful in distribution networks, as shown in Fig. 6.5, where there are large splitting losses or a large number of taps. Booster amplifiers are also needed when it is required to simultaneously amplify a number of input signals at different wavelengths, as is the case in WDM transmission. At high bit rates (typically > 2.5 Gb/s) directly modulated semiconductor lasers are susceptible to wavelength chirping where the laser oscillation wavelength changes with driving current. Chirping increases the effective laser spectral width. This increases fibre chromatic dispersion leading to an increase in ISI and degradation of the system BER. To avoid this effect at high bit rates lasers are usually externally modulated. The intrinsic insertion loss of external modulators can be compensated through the use of optical booster amplifiers.

*Table 6.2.* Applications of optical booster amplifiers

| |
|---|
| Increase medium-haul optical transmission link distance |
| Increase long-haul optical transmission link power budget |
| Compensate for splitting and tap losses in optical distribution networks |
| Simultaneous amplification of WDM signals |

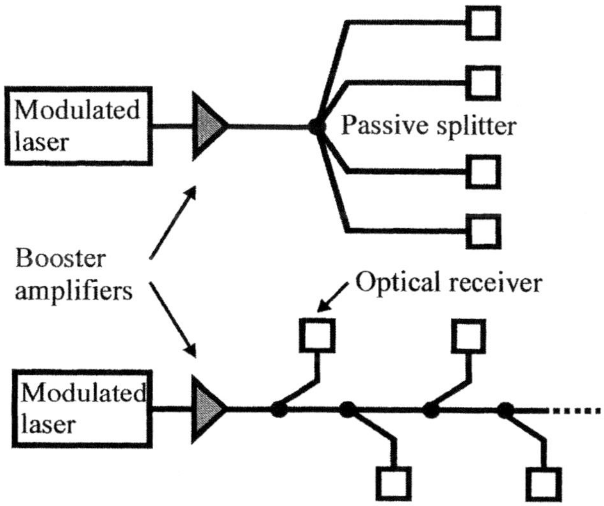


*Figure 6.5.* Booster amplifier application in optical distribution networks.

As the input signal power to a booster amplifier is usually high, the amplifier must have a high output saturation power $P_{o,sat}$. This enables a high amplifier output signal power to be achieved and also reduces pattern effects due to gain saturation. Pattern effects increase in severity as the amplifier input signal power and bit rate increase. A high $P_{o,sat}$ is also necessary for the application of a booster amplifier in WDM transmission systems [6]. In this case the saturation characteristics of the amplifier are determined by the total input power. This is due to the homogenous nature of the amplifier gain medium. A high $P_{o,sat}$ reduces interchannel crosstalk. A broad amplifier gain spectrum is also required for multi-wavelength channel amplification. SOA structures that can be used to realise high saturation output power are described in Chapter 3.

A typical booster SOA module is shown in Fig. 6.6 [7]. The module consists of a tapered waveguide MQW SOA chip mounted on a temperature controllable platform and coupling optics with aspheric lenses together with an output isolator, beam sampler and photodiode for output power control. Polarisation maintaining (PMF) and standard single-mode optical fibre are used for the input and output connections respectively.

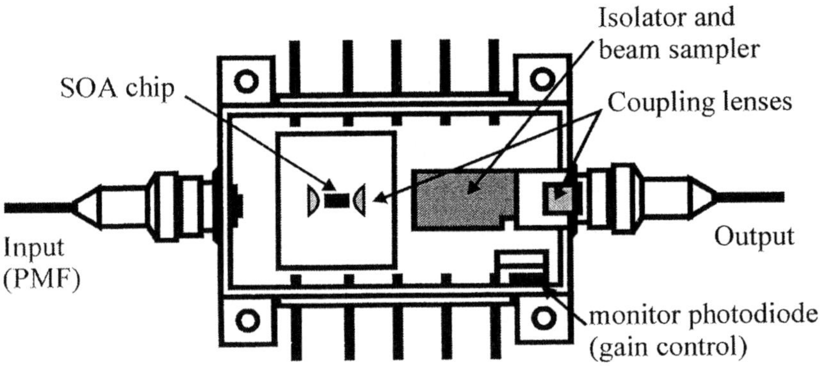


*Figure 6.6.* Schematic structure of an SOA booster module (after [7]).

An example of the use of an optical booster amplifier to increase a medium-haul transmission link distance is shown in Fig. 6.7 [8]. In this system trial the transmitter comprised a DFB laser directly modulated by a $2^{31}$-1 long pseudo-random bit sequence (PRBS) at 10 Gb/s. The laser emission wavelength was 1310 nm corresponding to the minimum material dispersion region of standard single-mode optical fibre. An optical booster amplifier module follows the laser. The SOA used in the module was an MQW device with four compressively strained quantum wells. The chip gain was 21 dB with a 3-dB saturation output power of 20 dBm. To minimise the effects of external reflections optical isolators were used on either side of the SOA. The isolators only pass a single polarisation state. The output isolator also halves the ASE from the amplifier. In the module the input light is focussed onto the isolator by a single aspheric lens giving a coupling loss of less than 2 dB. The module also contains a thermoelectric cooler and a thermistor to enable temperature control of the SOA chip. Experiments using the above laser and booster module with a p-i-n photodiode receiver gave a maximum power budget of approximately 26.8 dB for an average booster output amplifier power of 15.8 dBm. This power budget is enough to bridge 70 km of single-mode fibre. In the experimental trials it was found that the receiver sensitivity penalty caused by the booster amplifier is less than 1 dB if the average booster output power is kept below its 1-dB saturation output power of approximately 15 dBm. The maximum power budget is obtained at approximately 1 dB higher output power. The typical pulse shape distortion caused by dynamic gain saturation at this operating power is shown in Fig. 6.7. When an optical preamplifier was used before reception, the power budget was increased to 39.4 dB giving a potential unrepeatered span of 100 km at 10 Gb/s.

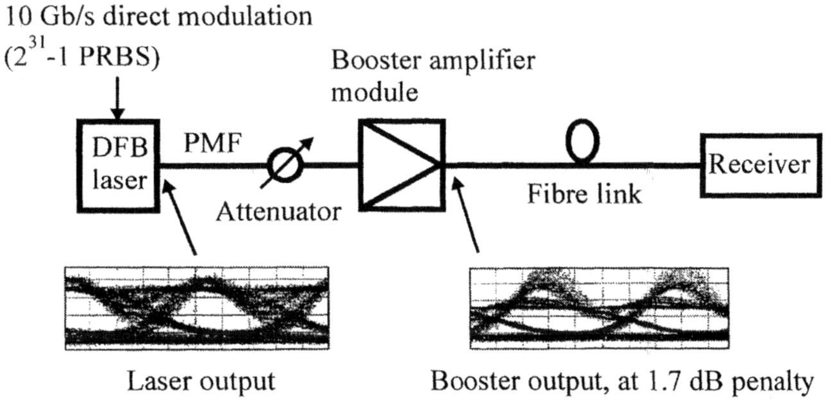


*Figure 6.7.* Booster amplifier transmission experiment with laser and booster output eye diagrams (after [8] with permission of the *IEEE*).

### 6.3.2 Preamplifier

The function of an optical preamplifier is to increase the power level of an incoming signal prior to conventional reception and demodulation. The increase in power level can greatly increase receiver sensitivity and so increase the link power budget. This allows longer unrepeatered links to be constructed. IM-DD receivers using optical preamplifiers are significantly more sensitive than conventional IM-DD receivers using p-i-n photodiodes or APDs. Compared to an APD, n optical amplifier delivers a larger gain and a wider bandwidth. There is no benefit in using an optical preamplifier in a coherent receiver as the local oscillator signal can be increased to a level where shot-noise limited performance is attained.

A schematic diagram of a preamplified digital optical receiver is shown in Fig. 6.8. The receiver consists of an optical preamplifier of uniform gain $G$, an optical filter of bandwidth $B_o$, a p-i-n photodiode with quantum efficiency $\eta$ followed by post-detection circuitry of electrical bandwidth $B_e$ and a decision circuit. The optical and electrical filters must be at least as wide as the modulated signal bandwidth to avoid signal distortion. The optical filter reduces the spontaneous emission reaching the detector leading to a reduction in the ASE dependent current noise. If the polarisation state of the incoming light signal is known, a polariser can be inserted between the preamplifier and the detector. The polariser passes the amplified signal but removes ASE in the orthogonal polarisation. This improves the receiver sensitivity by 3 dB in the signal-spontaneous beat noise limit. However, the polarisation state of the light signal is not usually known. The post-detection circuitry amplifies and processes the photocurrent from the detector. The most common type of post-detection processing is the 'integrate and dump' technique discussed in Section 6.1.1.

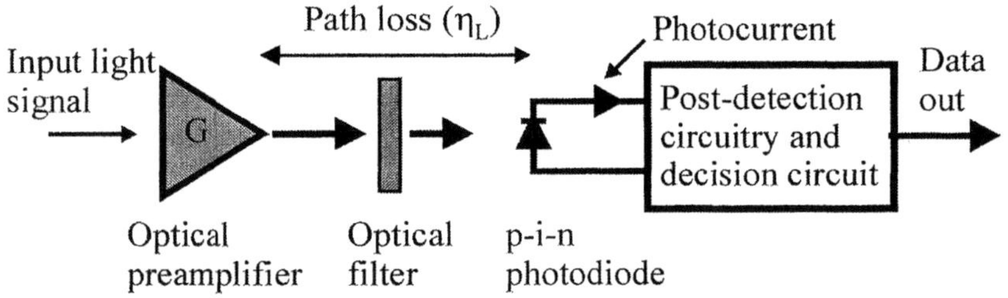


*Figure 6.8.* Preamplified optical receiver

In an optical receiver with no optical preamplifier the dominant photocurrent noise is circuit noise. This noise is typically one to two orders of magnitude greater than the receiver shot-noise limit. In a preamplified IM-DD optical receiver with a narrowband optical filter the dominant photocurrent noise is signal-spontaneous beat noise. Assuming that the amplifier gain G >> 1, the receiver signal-to-noise ratio (6.31) becomes

$$\left(\frac{S}{N}\right)_{out} = \frac{P_{sig}}{4n_{sp}h\nu B_e} \tag{6.36}$$

Typical improvements in the SNR compared to conventional receivers are greater than 10 dB. The form of (6.36) indicates that in the signal-spontaneous beat noise limit the receiver SNR is independent of both the amplifier gain and the optical filter bandwidth.

There is also a path loss $\eta_L$ between the preamplifier and detector. This is mainly caused by the optical filter loss and the coupling efficiency to the optical detector. This coupling loss will not lead to any degradation in the receiver performance as long as the product $G\eta$ >> 1. This consists favourably with a conventional receiver where the power penalty is equal to the coupling loss. More detailed analyses of optically preamplified receivers are given in [9-12].

POWER PENALTY DUE TO FINITE EXTINCTION RATIO

In practical systems the laser diode optical source is usually biased above threshold to ensure high-speed operation and reduce laser chirp. This means that some optical power will be present in the spaces of the transmitted optical bit stream, i.e. the extinction ratio is not infinite. This leads to a receiver power penalty. The power penalty in this case is defined as the difference in dB between the signal powers giving identical BERs for the finite and infinite extinction ratio $r$. The penalty is given [13] by

$$Penalty = \begin{cases} 10\log\left[\dfrac{(r+1)}{(r-1)}\right] & \text{circuit noise dominant} \\ 10\log\left[\dfrac{(r+1)(\sqrt{r}+1)}{(r-1)(\sqrt{r}-1)}\right] & \text{signal - spontaneous noise dominant} \end{cases} \tag{6.37}$$

The penalty is plotted in Fig. 6.9 as a function of $r$. A penalty < 1 dB requires $r$ > 10 dB in the case of a thermal noise limited receiver and $r$ > 20 dB in the case of a signal-spontaneous noise limited receiver. However the base sensitivity of the latter receiver is superior to the former. In an optically preamplified receiver, extinction ratio degradation is primarily caused by amplifier gain saturation.

A transmission experiment demonstrating the use of optical preamplifiers is shown in Fig. 6.10 [14]. The experiment uses the 1300 nm fibre window to obtain near dispersion free transmission at 10 Gb/s. The system requires a greater power budget than would be the case in the 1550 nm region due to the increased fibre loss. The transmitter laser is modulated with a 10 Gbit/s $2^7$-1 PSBS NRZ data pattern. The data had rise and fall times of 38 ns and 72 ns respectively with an extinction ratio of 7.8 dB. The optical receiver consists of a 50 Ω optical front-end and a.c. coupled broadband electrical amplifier. The electrical bandwidth of the amplifier was 10.6 GHz. This gave a sensitivity of -13.7 dBm at 10 Gbit/s for a BER of $10^{-9}$. To boost the receiver sensitivity a scheme of two cascaded 1310 nm MQW SOA preamplifiers with an optical bandpass filter, centred at the signal wavelength, prior to reception was used. The total cascaded gain of the two amplifiers was 26.8 dB with a polarisation sensitivity of 3.4 dB. The presence of an optical isolator between the two amplifiers improves the gain saturation performance compared to a single amplifier of the same gain. The second isolator prevents reflections from the optical filter coupling to the second SOA. Using a 0.2 nm bandwidth optical filter the best sensitivity obtained was -30.3 dBm. The transmission link was comprised of 89 km of conventional single-mode optical fibre with a total loss of 34 dB and a zero dispersion wavelength of 1305 nm. The signal wavelength was 1314.8 ns, giving rise to a small receiver penalty of 0.3 dB.

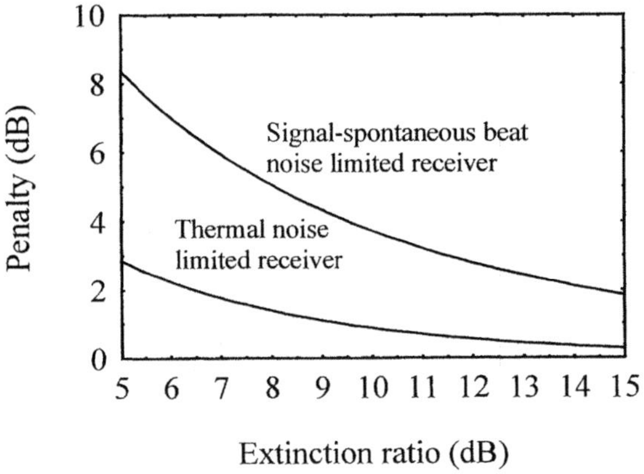


*Figure 6.9.* Optical receiver power penalty versus signal extinction ratio.

### 6.3.3 In-line amplifier and amplifier cascades

In loss limited optical communication systems, in-line optical amplifiers can be used to compensate for fibre loss thereby overcoming the need for optical regeneration. The main advantages of in-line SOAs are: Transparency to data rate and modulation format (in the unsaturated regime and at high bit rates), bidirectionality, WDM capability, simple mode of operation, low power consumption and compactness. The latter two advantages are particularly important for remotely located optical components.

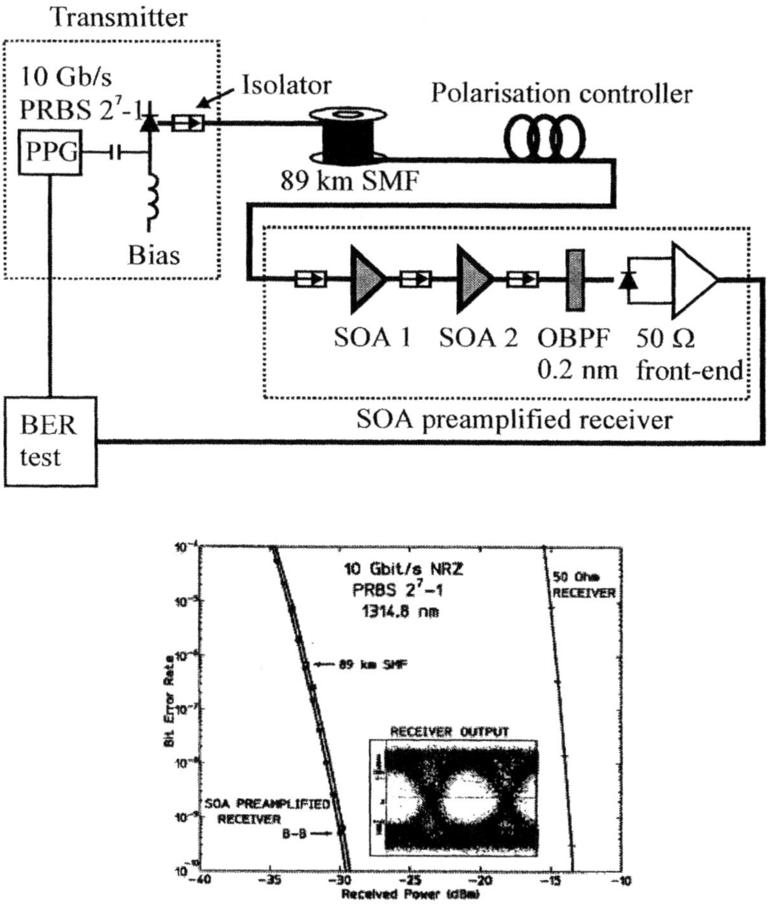


*Figure 6.10.* 89 km transmission experiment using two SOA preamplifiers and the BER performance of the 10 Gbit/s 1310 nm SOA preamplified receiver. The receiver eye diagram is shown for a received optical power of -29.8 dBm giving a BER of 6.3 x $10^{-9}$. OBPF: Optical bandpass filter, PPG: Pulse pattern generator (after [14] with permission of the *IEEE*).

### NOISE FIGURE OF AN AMPLIFIER CASCADE

A good figure of merit for an optical amplifier cascade is the overall noise figure $F_{tot}$ at the output of the last amplifier in the link [15]. To evaluate $F_{tot}$ consider the amplifier cascade of Fig. 6.11, which consists of $k$ amplifiers of gain $G_i$ (including coupling loss) and noise figure $F_i$ ($i = 1 \ldots k$) separated by fibre links of loss $L_i$. Furthermore it is assumed that a narrowband filter centred at the signal wavelength is present at each amplifier output so the system operates in the signal-spontaneous beat noise limit. Depending on the length of the final link, the actual receiver SNR may be signal-spontaneous beat noise or receiver circuit noise limited. Each amplifier is assumed to operate in the non-saturated regime.

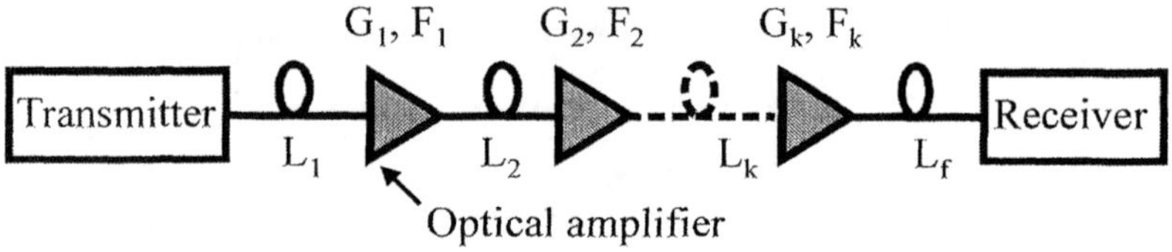


*Figure 6.11.* Optical fibre link with cascaded amplifiers.

The total noise figure, after the *k*-th amplifier is

$$\begin{aligned} F_{tot} &= \frac{(S/N)_{in}}{(S/N)_{out}} \\ &= \frac{F_1}{L_1} + \frac{F_2}{L_1 G_1 L_2} + \ldots + \frac{F_k}{\left(\sum_{i=1}^{k} L_i G_i\right) L_k} \end{aligned} \tag{6.38}$$

where $(S/N)_{in}$ and $(S/N)_{out}$ are the input and output SNRs of the amplifier cascade respectively. If we consider the case where $L_i = 1$, i.e. a composite amplifier of overall signal gain $G_{tot} = kG$, constructed of a chain of smaller gain amplifiers of gain $G$ separated by narrowband optical filters, then

$$F_{tot} = F_1 + \frac{F_2}{G} + \ldots + \frac{F_k}{G^k} \tag{6.39}$$

This expression indicates that a low-noise, high gain amplifier can be constructed by combining a low noise first stage amplifier followed by high saturation output power amplifiers. This technique is particularly useful when constructing low noise optical preamplifiers as in the above transmission experiment where the isolator between the two SOAs acts to improve the noise figure of the first amplifier and increase the saturation output power of the second.

If we next consider the case where the function of each amplifier in the cascade is to compensate for the losses of the preceding fibre, i.e. $G_i L_i = 1$, then from (6.38) the total noise figure of the cascade is

$$F_{tot} = \sum_{i=1}^{k} G_i F_i \tag{6.40}$$

If all the $G_i$ and $L_i$ are equal then

$$F_{tot} = kGF \tag{6.41}$$

This implies that each amplifier contributes equally to the total noise figure of the system. If the allowable SNR at the receiver is known, then the maximum number of cascadable amplifiers $k_{max}$ can be determined. The maximum unrepeatered link distance is equal to $k_{max}l$ where $l$ is the distance between successive amplifiers. To maximise the link distance it is important to have low-noise, high saturation output power amplifiers.

In a link with no in-line amplifiers the SNR as a function of distance $z$ from the transmitter is [16],

$$SNR(z) = \frac{P_s^2 e^{2\alpha_f z}}{\left(2eB_e P_s e^{\alpha_f z} + \dfrac{4kTF_e B_e}{R_L}\right)} \tag{6.42}$$

In (6.42) it is assumed that the detector has unity quantum efficiency. The first and second terms in the denominator of (6.42) are signal shot-noise and receiver thermal noise respectively. $F_e$ is the receiver noise figure, $B_e$ the electrical bandwidth, $R_L$ the photodetector load impedance and $\alpha_f$ the fibre loss coefficient. Fig. 6.12 shows the SNR of a practical optically amplified link as a function of distance compared to that of a non-amplified link. The SNR of an optically amplified link decreases very slowly with distance. In contrast the SNR of a non-amplified link degrades rapidly. In this case the maximum distance that can be traversed by an non-amplified link, such that the BER at the receiver is $< 10^{-9}$ for NRZ data, is 145 km. For the amplified link, if SNR is the only factor taken into account, the overall link distance possible is essentially infinite. In practise other factors such as amplifier saturation caused by ASE, fibre dispersion (especially chromatic dispersion), polarisation effects and fibre nonlinearities limit the distance possible between regenerators.

NOISE SATURATION IN AN AMPLIFIER CHAIN

A major design consideration in amplifier chains is gain saturation caused by accumulation of amplifier noise. Amplifier saturation causes a closure in the system eye diagram leading to an increase in the BER. We consider the amplifier chain of Fig. 6.11 with $k$ identical amplifiers of gain $G$ separated by fibre lengths of loss $L$, where $GL = 1$. In addition it is also assumed that optical filters are present at the output of each amplifier. The filters are assumed to have an ideal rectangular response centred at the signal wavelength. The total output noise power from the k-th amplifier is

$$N_k = 2kn_{sp}h\nu\eta_{out}\left(\frac{G}{\eta_{in}\eta_{out}} - 1\right)B_o \tag{6.43}$$

where $\eta_{in}$ and $\eta_{out}$ are the amplifier input and output coupling losses respectively. A similar formula applies to noise propagation in the opposite direction. This backward propagating noise can be eliminated using isolators. However the use of isolators does not allow bi-directional transmission and also precludes the use of fault detection techniques such as optical time domain reflectometry.

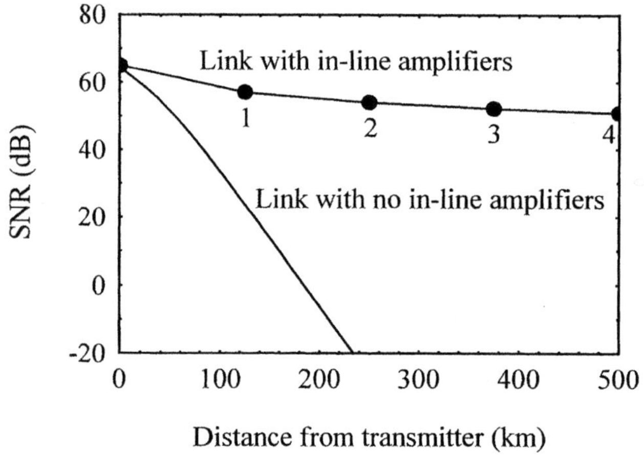


*Figure 6.12.* SNR versus distance in a fibre link. The link parameters are: 0.2 dB/km fibre attenuation, 1 mW launched transmitter signal power, 25 dB amplifier gain and 8 dB amplifier noise figure. The receiver parameters are: Load impedance of 1.0 kΩ, temperature of 300 K and an electrical bandwidth of 1 GHz. In the optically amplified link case the signal and ASE are passed through a narrowband optical filter prior to reception. The amplifiers are spaced at intervals such the amplifier gain compensates exactly for the preceding fibre loss.

The noise can also be greatly reduced using narrowband optical filters. However, if the filter bandwidth is too narrow the system optical bandwidth is limited so preventing capacity enhancements using WDM technology.

To avoid serious degradation in system performance the total output noise power from each amplifier in the chain must be less than the amplifier saturation output power $P_{o,sat}$. Hence the maximum number of amplifiers allowed in the link, from gain saturation considerations, is

$$k_{\max} = \frac{P_{o,sat}}{2n_{sp}h\nu\eta_{out}\left(\dfrac{G}{\eta_{in}\eta_{out}} - 1\right)B_o} \tag{6.44}$$

As an example consider a cascade of SOAs operating in the 1550 nm region with the following parameters: 25 dB fibre-to-fibre gain, 3 dB input and output coupling losses and 5 dBm saturation output power. If a 10 nm bandwidth optical bandpass filter is used at the output of each amplifier, and if it is assumed that saturation caused by the transmitted signal is negligible, then using (6.44) the maximum number of cascadable amplifiers is fifteen. The precise influence of SOA noise on system performance depends on various parameters such as extinction ratio, chirp, modulation format, bit rate and receiver design.

A single channel transmission experiment illustrating the use of in-line SOAs is shown in Fig. 6.13 [17]. The transmitter comprises a 1309 nm DFB laser directly modulated with a 10 GHz frequency generator to produce a train of pulses with a full width at half maximum of 40 ps. At 1309 nm fibre chromatic dispersion is small and the maximum transmission distance is mainly limited by fibre attenuation. The laser output is connected to a $LiNiO_3$ modulator. The electrical input to the modulator is a $2^{31}$-1 PRBS that produces an RZ optical data stream with an extinction ratio of 13 dB. A booster MQW SOA is used to increase the average transmitted power to between 0 and 2 dBm. The transmission length is 420 km with twelve in-line MQW SOAs, used to compensate for fibre loss, spaced at 38 km intervals. Polarisation independent isolators were used at the output of each amplifier. Optical bandpass filters were not used along the fibre link. At the receiver the signal was amplified by an optical preamplifier, identical to the in-line amplifiers, and filtered by a 1 nm bandpass optical filter. The signal was then detected by a PIN photodiode followed by an electronic clock and data recovery circuit. The optical preamplifier and filter increase the receiver sensitivity from -14 dBm to -31 dBm. The receiver penalty after 420 km was 5 dB. In this experiment, the main limitation on transmission distance was the accumulation of ASE within the optical filter bandwidth.

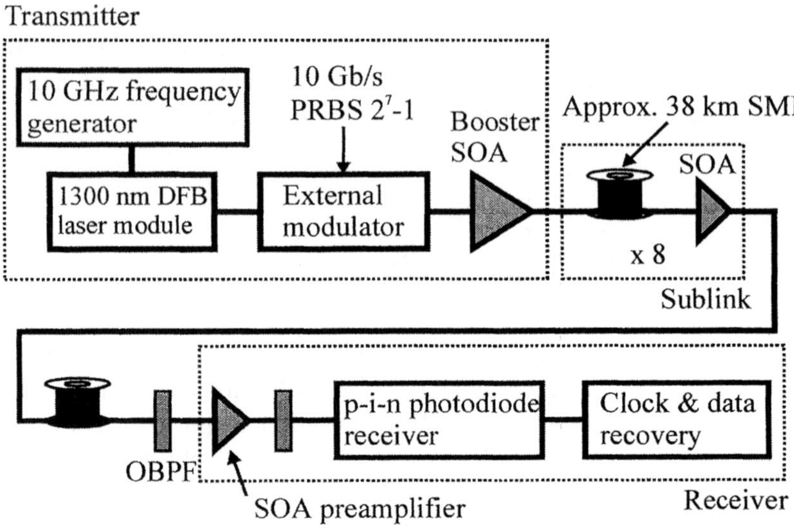


*Figure 6.13.* 10 Gbit/s, RZ transmission experiment over 420 km at 1309 nm. The in-line SOAs are spaced at 38 km intervals. OBPF: Optical bandpass filter (after [17] with permission of the author).

### WDM TRANSMISSION

WDM transmission can be used to greatly enhance the capacity of optical fibre links. As SOAs have wide gain bandwidths, they can be utilised to simultaneously amplify many wavelengths. To avoid amplifier gain saturation, the total input signal power must be less than the saturation input power $P_{i,sat}$, i.e.

$$\sum_{k=1}^{N} P_{k,in} < P_{i,sat} \tag{6.45}$$

where $N$ is the number of transmitted wavelengths and $P_{k,in}$ is the amplifier input signal power at wavelength $\lambda_k$.

An example of a WDM transmission experiment is shown in Fig. 6.14 [18]. The transmitter consists of eight external cavity lasers combined by an 8:1 coupler. The wavelengths are in the range 1558-1570 nm with a channel spacing of 200 GHz. The channels are externally modulated at 20 Gbit/s by a $2^{13}$-1 PRBS and passed through a 2 km length of single mode fibre to produce independent bit patterns. Three booster SOAs (bulk-tensile strained InGaAsP active region devices with saturation output powers of 12 dBm) are used to compensate for the coupler and modulator losses. The transmission link is comprised of four amplified 40 km single-mode fibre links including dispersion compensating fibre. The span loss is 13 dB. The gain peak of each in-line amplifier is at 1530 nm. The 12 to 14 dB of gain available from each amplifier is adequate to compensate for the link loss. The receiver consists of two SOA preamplifiers between which the signal is demultiplexed to 10 Gbit/s by an $LiNiO_3$ modulator. The demultiplexed data is then detected by a p-i-n diode. Using pre-emphasis an optical SNR of 24 dBm (in a 0.1 nm resolution bandwidth) was obtained for all channels. All the channels had detected *Q*-factors between 17 and 18 dB, corresponding to a BER < 3 x $10^{-13}$. Fig. 6.15 shows the dependence of the *Q*-factor on the total launched power. The SNR is degraded at low launched powers due to ASE and at high launched powers by ISI and interchannel crosstalk caused by self-gain modulation (channel pattern effects) and interchannel crosstalk.

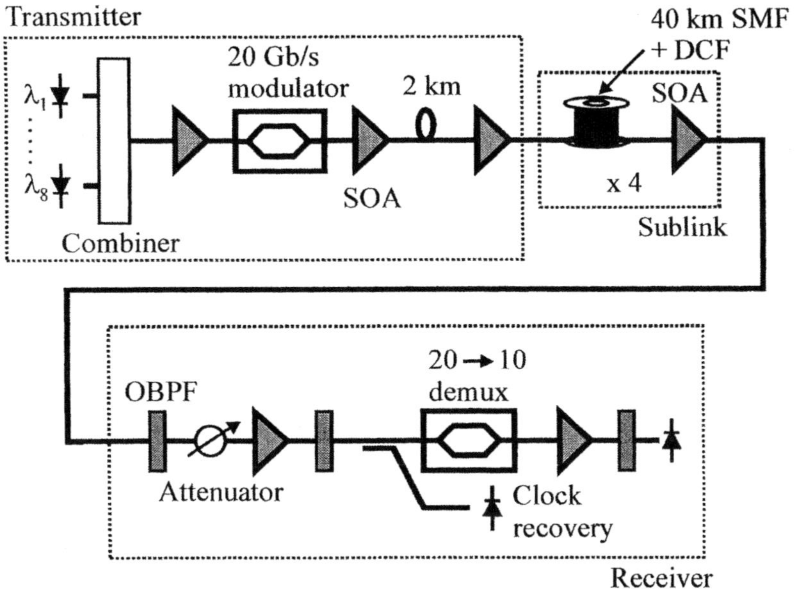


*Figure 6.14.* 8-channel DWDM transmission experiment. DCF: Dispersion compensating fibre (after [18] with permission of the *IEEE*).

## REDUCTION OF INTERCHANNEL CROSSTALK

Crosstalk can be avoided by operating an SOA in the small signal region. However, amplifiers need to operate in the saturation regime in order to achieve high output power and efficiency. Various techniques have been investigated to reduce interchannel crosstalk [19-23]. The main objective of these methods is to minimise changes in the amplifier carrier density. Gain clamped SOAs (Chapter 3) can be used for this purpose [19-20].

In [21] a polarising multiplexing technique is used to reduce WDM crosstalk in SOAs. In this method, shown in Fig. 6.16(a), two laser sources are split into two streams, which are then modulated by the data signal and its complement. The polarisations of the signals are adjusted to be mutually

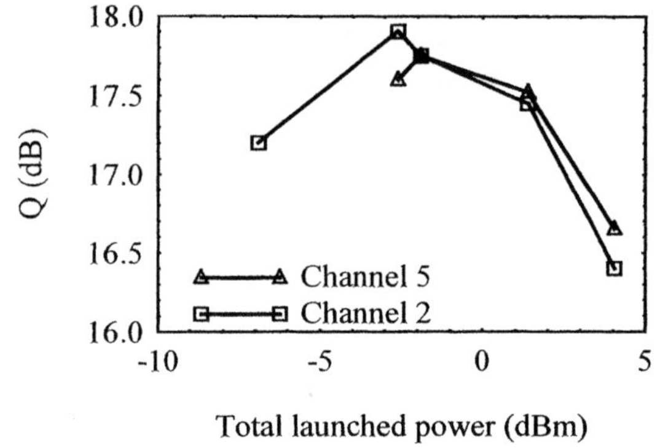


*Figure 6.15.* *Q*-factor versus total launched power (including ASE) at the end of the 160 km transmission link of Fig. 6.14 (after [18] with permission of the *IEEE*).

orthogonal. The signals are then multiplexed with a suitable delay such that the vector sum of the two streams has a constant value while being amplified. At the receiver end the individual polarisation streams are demultiplexed and the power of one of the states is optimised for detection and data recovery. The technique leads to a large improvement in system performance as shown by the eye diagrams of Fig 6.16(b).

Another technique that has been used to reduce crosstalk is feed forwarding [22]. In this method the SOA carrier density is maintained at a constant value by driving the device with a constant bias current and a signal proportional to the appropriately delayed and weighted sum of the electrical drive currents of the signal source laser diodes. The above techniques are relatively complex and not cost effective.

In [23] transmission of 32 DWDM channels through three in-line SOAs was achieved without local gain control. This was made possible due to three factors: (a) Optimisation of the receiver sensitivity by adjustment of the receiver detection threshold. (b) Fluctuations in the total input power to the amplifier decrease as the number of transmitted wavelength channels increases. This is because the data streams carried on each channel are uncorrelated, so the amplifier carrier density only depends on the slowly changing total input signal power. (c) An external reservoir channel (unmodulated) is transmitted along with the modulated channels to further suppress the effect of power fluctuations.

### 6.3.4 SOAs in analog transmission

In general analog transmission systems have much tighter constraints compared to digital systems, which has limited the application of SOAs in these networks [24]. As is the case for digital transmission, degrading effects such as signal distortion and intermodulation distortion can be reduced by operating SOAs in the unsaturated regime [25], employing gain clamped devices [26] or using gain control techniques such as those discussed above.

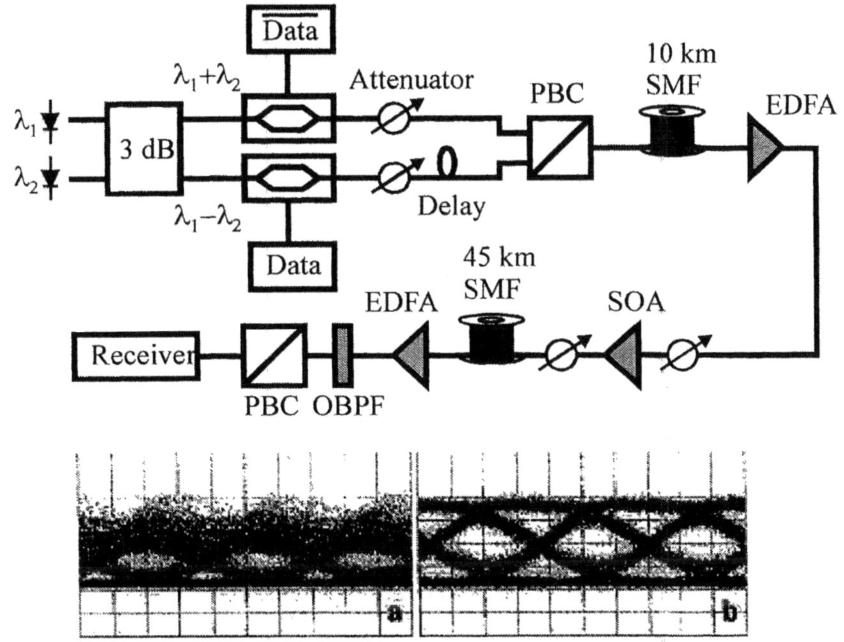


*Figure 6.16.* Top: Experimental setup for the polarisation multiplexing technique used to reduce crosstalk in SOAs. Bottom: Eye-diagrams after amplification by the SOA with (left) and without (right) polarisation multiplexing. PBC: Polarisation beam combiner (after [20] with permission of the *IEEE*).

# REFERENCES

1. J. M. Senior, *Optical Fiber Communications: Principles and Practice* - 2nd ed., Prentice Hall (1992).
2. L. Kazovsky, S. Benedetto and A. Willner, *Optical Fiber Communication Systems*, Artech House (1996).
3. S. Ryu, *Coherent Lightwave Communication Systems*, Artech House (1995).
4. G. Jacobsen, *Noise in Digital Optical Transmission Systems*, Artech House (1994).
5. N. A. Olsson, Lightwave systems with optical amplifiers, *J. Lightwave Technol.*, **7**, 1071-1082 (1989).
6. M. Koga and T. Matsumoto, The performance of a travelling-wave type semiconductor optical amplifier as a booster in multiwavelength simultaneous amplification, *J. Lightwave Technol.*, **8**, 105-112 (1990).
7. S. Tsuji, T. Toyonaka, N. Kikuchi, S. Aoki and S. Sasaki, High-power (+15dBm) semiconductor optical booster module for high-bit-rate transmission systems, Paper FG5, *Optical Fiber Communications Conference*, OFC'94 (1994).
8. L.F. Tiemeijer, P.J.A. Thijs, T. Van Dongen, J.J.M. Binsma, E.J. Jansen, P.K. Kuindersma, G.P.J.M. Cuijpers and S. Walczyk, High-output-power (+15 dBm) unidirectional 1310-nm multiple-quantum-well booster amplifier module, *IEEE Photon. Technol. Lett.*, **12**, 1519-1521 (1995).
9. D. Marcuse, Derivation of analytical expressions for the bit-error probability in lightwave systems with optical amplifiers, *J. Lightwave Technol.*, **8** 1816-1823 (1990).
10. D. Marcuse, Calculation of bit-error probability for a lightwave system with optical amplifiers and post-detection Gaussian noise, *J. Lightwave Technol.*, **9** 505-513 (1991).
11. M.S. Leeson, Fast approximation for bit error rate calculations in optically preamplified receivers, *Electron. Lett.*, **33** 1329-1330 (1997).
12. L.G. Kazovsky and O.K. Tonguz, Sensitivity of direct-detection lightwave receivers using optical preamplifiers, *Photon. Technol. Lett.*, **3**, 53 -55 (1991).
13. A.J. McDonald, R.S. Fyath and J.J. O'Reilly, Influence of extinction ratio on performance of optical receivers incorporating laser preamplifiers, *Electron. Lett.*, **25**, 249-250 (1989).
14. H. de Waardt, L. F. Tiemeijer, B. H. Verbeek, 89 km 10 Gbit/s 1310 nm repeaterless transmission experiments using direct laser modulation and two SL-MQW laser preamplifiers with low polarization sensitivity, *IEEE Photon. Technol. Lett.*, **6**, 645-647 (1994).
15. Y. Yamamoto, Ed., *Coherence, Amplification, and Quantum Effects in Semiconductor Lasers*, John Wiley (1991).
16. A. Yariv Signal-to-noise considerations in fibre links with periodic or distributed amplification, *Optics Lett.*, **15**, 1064-1066 (1990).
17. P.I. Kuindersma, G.P.J.M. Cuijpers, J.G.L. Jennen, J.J.E. Reid, L.F. Tiemeijer, H. de Waardt and A.J. Boot, 10 Gbit/s RZ transmission at 1309 nm over 420 Km using a chain of multiple quantum well semiconductor optical amplifier modules at 38 km intervals, *Proc. 22nd European Conference on Optical Communication*, ECOC '96, **2**, 165-168 (1996).
18. L.H. Spiekman, J.M. Wiesenfeld, A.H. Gnauck, L.D. Garrett, G.N. Van Den Hoven, T. Van Dongen, M.J.H. Sander-Jochem and J.J.M. Binsma, Transmission of 8 DWDM channels at 20 Gb/s over 160 km of standard fiber using a cascade of semiconductor optical amplifiers, *IEEE Photon. Tech. Lett.*, **12**, 717-719 (2000).

19. Sun Jinying, G. Morthier and R. Baets, Numerical and theoretical study of the crosstalk in gain clamped semiconductor optical amplifiers, *IEEE J. on Sel. Topics in Quantum Electron.*, **3**, 1162-1167 (1997).
20. D.T. Schaafsma and E.M. Bradley, Cross-gain modulation and frequency conversion crosstalk effects in 1550-nm gain-clamped semiconductor optical amplifiers, *IEEE Photon. Technol. Lett.*, **11**, 727-729 (1999).
21. A.K Srivastava, S Banerjee, B.R. Eichenbaum, C. Wolf, Y. Sun, J.W. Sulhoff and A.R. Chraplyvy, A polarization multiplexing technique to mitigate WDM crosstalk in SOAs, *IEEE Photon. Tech. Lett.*, **12**, 1415-1416 (2000).
22. C.R. Doerr, C.H. Joyner, M. Zirngibl, L.W. Stulz and H.M. Presby, Elimination of signal distortion and crosstalk from carrier density changes in the shared semiconductor amplifier of multifrequency signal sources, *IEEE Photon. Technol. Lett.*, **7**, 1131-1133 (1995).
23. Y. Sun, A.K. Srivastava, S. Banerjee, J.W. Sulhoff, R. Pan, K. Kantor, R.M. Jopson and A.R. Chraplyvy, Error-free transmission of 32 x 2.5 Gbit/s DWDM channels over 125 km using cascaded in-line semiconductor optical amplifiers, *Electron. Lett.*, **35**, 1863-1865 (1999).
24. T.E. Darcie, and G.E. Bodeep, Lightwave subcarrier CATV transmission systems, *IEEE Trans. Microwave Theory and Techniques*, **38**, 524-533 (1990).
25. K.D. LaViolett, The use of semiconductor-optical-amplifiers for long optical links in the CATV upstream optical network, *IEEE Photon. Technol. Lett.*, **10**, 1165-1167 (1998).
26. L.F. Tiemeijer, G. N. van den Hoven, P. J. A. Thijs, T. van Dongen, J. J. M. Binsma and E. J. Jansen, 1310 nm DBR type MQW gain clamped SOAs with AM-CATV grade linearity, *IEEE Photon. Technol. Lett.*, **8**, 1543-1545 (1996).

Chapter 7

# FUNCTIONAL APPLICATIONS

In addition to the basic applications described in the preceding chapter, SOAs can also be used to perform functions that are particularly useful in optically transparent networks. These all-optical functions can help to overcome the 'electronic bottleneck', where signal processing is carried out in the electrical domain. This is a major limiting factor in the deployment of high-speed optical communication networks. Many of these functional applications are based on SOA nonlinearities. The development of photonic integrated circuits (PICs) has made the deployment of relatively complex SOA functional elements more feasible.

In this chapter we first begin with an overview of SOA linearities. We then consider some of the many functional applications possible with SOAs, including their PIC implementation.

## 7.1 SOA NONLINEARITIES

Nonlinearities in SOAs are principally caused by carrier density changes induced by the amplifier input signals. The four main types of nonlinearity are: Cross gain modulation (XGM), cross phase modulation (XPM), self-phase modulation (SPM) and four-wave mixing (FWM).

### 7.1.1 Cross-gain modulation

The material gain spectrum of an SOA is homogenously broadened. This means that carrier density changes in the amplifier will affect all of the input signals. The carrier density temporal response is dependent on the carrier lifetime. As discussed in the preceding chapter, carrier density changes can give rise to pattern effects and interchannel crosstalk in multiwavelength amplification. The most basic XGM scenario is shown in Fig. 7.1 where a weak CW probe light and a strong pump light, with a small-signal harmonic modulation at angular frequency $\omega$, are injected into an SOA. XGM in the amplifier will impose the pump modulation on the probe. This means that the amplifier is acting as a wavelength converter, i.e. transposing information at one wavelength to another signal at a different wavelength.

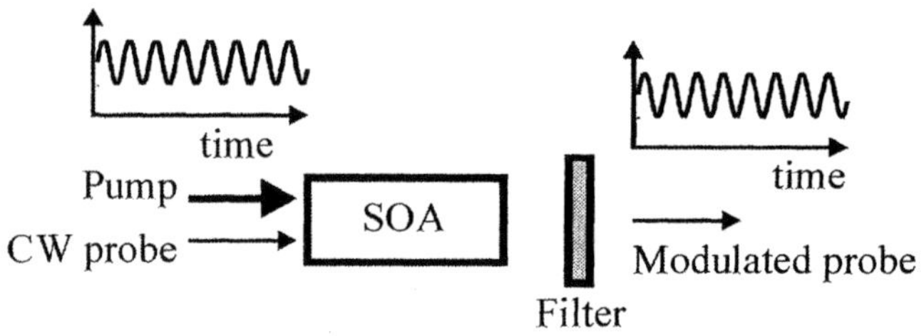


*Figure 7.1.* Simple wavelength converter using XGM in an SOA

The most useful figure of merit of the converter is the conversion efficiency $\eta$, which is defined as the ratio between the modulation index of the output probe to the modulation index of the input pump. In [1,2] a small-signal harmonic analysis is used to determine $\eta$, giving

$$\eta = \left| \frac{P_1(0)}{P_0(0)} F(L) \right| \tag{7.1}$$

with

$$\begin{aligned} F(L) &= 1 - e^{-K(L)} \\ K(L) &= \frac{1}{1 + j\omega\tau\alpha'} \left\{ \alpha' \ln \frac{G_0}{G} - \ln \left[ 1 - \frac{(G-1) P_T(0)/P_{sat}}{1 + GP_T(0)/P_{sat} + j\omega\tau} \right] \right\} \end{aligned} \tag{7.2}$$

where $\alpha' = \alpha/(\Gamma g_0)$ is the normalised waveguide loss coefficient. $P_0(0)$ and $P_1(0)$ are the probe and pump average input powers respectively, $P_T(0)$ the sum of the average input pump and probe powers, $\omega$ the small-signal harmonic modulation angular frequency and $\tau$ the spontaneous carrier lifetime. $g_0$ is given by (3.22). $G_0$, the unsaturated gain, is given by

$$G_0 = \exp[(\Gamma g_0 - \alpha)L] \tag{7.3}$$

The amplifier gain $G$ can be determined from

$$\alpha' \ln \frac{G_0}{G} = \ln\left\{\frac{1-\alpha'[1+P_T(0)/P_{sat}]}{1-\alpha'[1+G\,P_T(0)/P_{sat}]}\right\} \tag{7.4}$$

(7.4) is identical to (3.24) in the limit as $\alpha$ tends to 0. The above analysis assumes that the optical frequency difference (detuning) between the pump and probe signals is much greater than $1/(2\pi\tau)$ so four-wave mixing products produced by the nonlinear interaction between the two fields can be neglected. A typical response characteristic, computed using (7.2) and the numerical solution of (7.4), is shown in Fig. 7.2. The values used in the

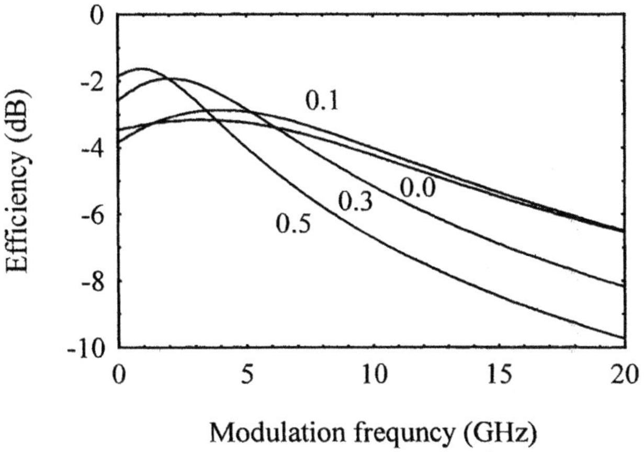


*Figure 7.2.* Typical frequency response of an SOA wavelength converter using XGM. The parameter is the normalised waveguide loss coefficient (after [1] with permission of the IEEE).

calculations are $\exp(\Gamma g_0 L) = 30$ dB, $P_0(0) = 0.5$ mW, $P_1(0) = 1.0$ mW, $P_{sat} = 10$ mW and $\tau = 0.1$ ns. Fig. 7.2 shows that the response bandwidth is very sensitive to waveguide losses.

In the case of zero waveguide loss, we have

$$F(L)_{\alpha \to 0} = \frac{(G-1)\,P_T(0)/P_{sat}}{1+G\,P_T(0)/P_{sat} + j\omega\tau} \tag{7.5}$$

In this case the converter response has a low-pass filter characteristic with 3-dB bandwidth

$$f_{3dB} = \frac{1}{2\pi}\left[\frac{1 + G\,P_T(0)/P_{sat}}{\tau}\right] \tag{7.6}$$

### 7.1.2 Self-phase and cross-phase modulation

The refractive index of an SOA active region is not constant but is dependent on the carrier density and so the material gain. This implies that the phase and gain of an optical wave propagating through the amplifier are coupled via gain saturation. This strength of this coupling is related to the linewidth enhancement factor $\alpha_l$ (so termed because of its influence on the spectral linewidth of semiconductor lasers) of the material given by

$$\alpha_l = -\frac{\pi}{\lambda_0}\frac{dn_e/dn}{dg_m/dn} \tag{7.7}$$

where $\lambda_0$ is the free-space wavelength, $n_e$ the amplifier waveguide effective index, $g_m$ the material gain coefficient and $n$ the carrier density [3].Plots of $\alpha_l$ versus $\lambda_0$ with carrier density as parameter are shown in Fig. 7.3.

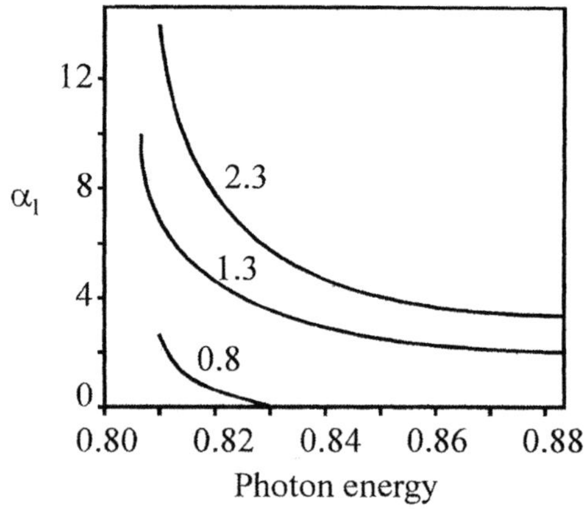


Figure 7.3. Calculated linewidth enhancement factor versus wavelength for undoped InGaAsP. The parameter is carrier density ($\times 10^{24}$ $m^{-3}$) (after [3] with permission of the IEEE).

$\alpha_l$ increases rapidly as the photon energy nears the bandgap energy due the sensitivity of the material gain coefficient to carrier density in this region. $\alpha_l$ is also a function of carrier density.

When an injected signal pulse travels through an SOA it causes carrier density changes, so changing its propagation coefficient (via the effective index). Because of the finite carrier lifetime, the leading edge of the pulse experiences a different phase shift relative to the lagging edge. This self-phase modulation (SPM) will change the pulse shape as well as its spectrum. This effect can be used to create a dispersion compensator where SPM pulse narrowing is used to compensate for pulse broadening resulting from chromatic dispersion in optical fibre. This type of device has great potential for increasing the capacity of high-speed long-haul fibre links.

If more than one signal is injected into an SOA, there will be cross-phase modulation (XPM) between the signals. XPM can be used to create wavelength converters and other functional devices. However, because XPM only causes phase changes, the SOA must be placed in some form of interferometric configuration to convert phase changes in the signals to intensity changes using constructive or destructive interference. Three common interferometers are shown in Fig. 7.4 [4].

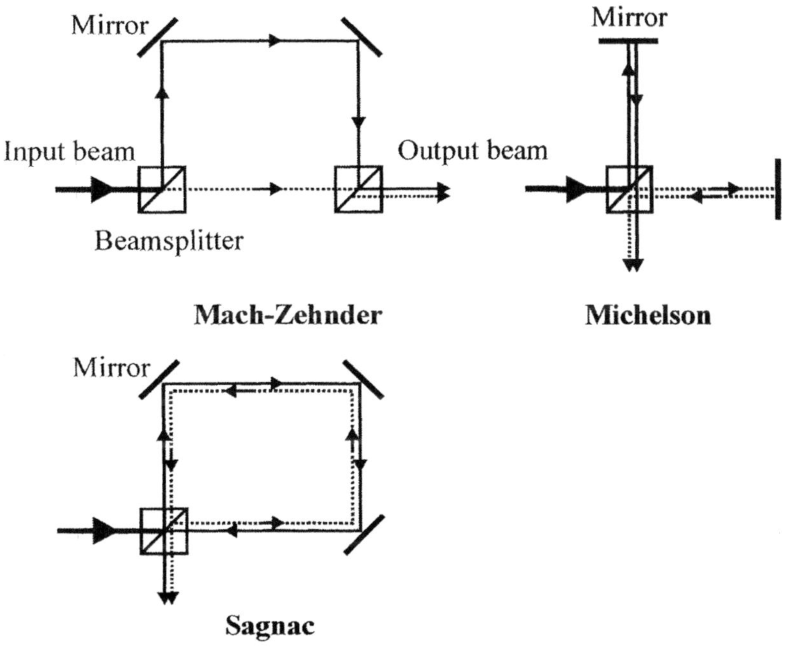


*Figure 7.4.* Common interferometers used in SOA functional applications

In the Mach-Zehnder interferometer (MZI), an input beam is split into two beams that traverse different paths before they are recombined at the interferometer output. Depending on the path difference, the returning beams will interfere constructively or destructively. The Michelson interferometer (MI) is similar to the MZI but requires only one beamsplitter. In the Sagnac interferometer the input beam is split into two beams that traverse the same distance but in opposite directions, before recombination. In interferometric configurations using SOAs, amplifiers are placed in one or more of the interferometer arms in order to actively change the phase of the propagating signal.

### 7.1.4 Cross-polarisation modulation

Most commercial SOAs have good polarisation independence. However, there is still some device birefringence due to the difference between the amplifier TE and TM mode effective indices. A very small index difference (e.g. $2 \times 10^{-4}$ for a 2 mm long device at 1550 nm) can cause a large TE-TM relative phase shift. This cross-polarisation modulation (XPOM) can affect the performance of interferometric or four-wave mixing structures using SOAs [5]. However it can be used in demultiplexing and wavelength conversion applications.

### 7.1.4 Four-wave mixing

Four-wave mixing (FWM) is a *coherent* nonlinear process that can occur in an SOA between two optical fields, a strong pump at angular frequency $\omega_0$ and a weaker signal (or probe) at $\omega_0$ - $\Omega$, having the same polarisation. The injected fields cause the amplifier gain to be modulated at the beat frequency $\Omega$. This gain modulation in turn gives rise to a new field at $\omega_0 + \Omega$, as shown in Fig. 7.5. The new field is called the conjugate because its phase is the opposite of the signal phase. This means that the spectrum of the conjugate signal is a shifted and inverted replica of the input signal. FWM generated in SOAs can be used in a many applications including wavelength converters, dispersion compensators and optical demultiplexers.

FWM in SOAs arises from different physical phenomena. At low $\Omega$, the dominant mechanism is modulation of the carrier density resulting from pump-signal beating. This is an *interband* effect as it involves carrier-hole recombination between the material conduction and valence bands. Because of this the characteristic time of this process is the carrier lifetime. This is the order of hundreds of picoseconds. This implies that this particular mechanism will only manifest itself for detuning frequencies of the order of tens of GHz.

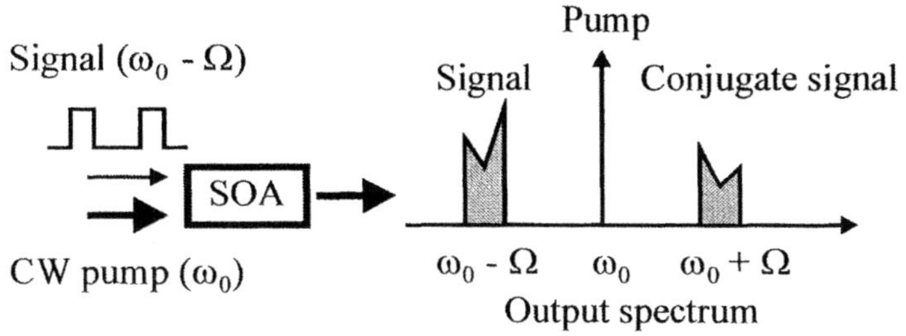


*Figure 7.5.* SOA FWM. The signal and pump must have identical polarisation states for efficient FWM to occur. For clarity the amplifier output ASE spectrum has been omitted.

There are two additional mechanisms that can cause FWM in SOAs: Spectral hole burning (SHB) and carrier heating (CH). SHB is caused by the injected pump signal creating a hole in the *intraband* carrier distribution. This effectively modulates the occupation probability of carriers within a band leading to fast gain modulation.

CH is caused by stimulated emission and free carrier absorption. Stimulated emission subtracts carriers that are cooler than average while free carrier absorption moves carriers to higher energy levels in the band. The resulting increase in temperature decreases the gain. There are two characteristic times associated with CH. The first is the carrier-phonon scattering time $\tau_1$, which is the average time carriers require to cool down to the lattice temperature. The second is the carrier-carrier scattering time $\tau_2$, which is the average time taken by the carrier population to reach a heated equilibrium from the initial non-heated equilibrium. Both SHB and CH have characteristic times of the order of hundreds of femtoseconds.

The analysis of FWM in SOAs is usually based on the coupled-mode equations [6]. These equations are complex and not amenable to analytic solution. However, if it is assumed that the power in the conjugate signal is small relative to the pump and probe signal then it is possible to obtain expressions for the normalised conjugate output $\rho$ and the efficiency $\eta$ [7]. $\rho$, the power of the conjugate normalised to the output power of the probe, is given by

$$\begin{aligned}\rho &\equiv \frac{P_2(L)}{P_1(L)} \\ &= |G'|^2 \exp\{\mathrm{Re}[(1-j\alpha_l)\sigma F_{cd}(L,-\Omega)]\}\end{aligned} \tag{7.8}$$

$\eta$, the ratio between the conjugate output power and the input probe power, is given by

$$\eta = G|G'|^2 \tag{7.9}$$

where

$$\begin{aligned}G' &= -\frac{1-j\alpha_l}{\alpha}\exp\left[-\frac{1}{2}\sigma F_{cd}(\Omega)\right]\sin\left[\frac{\alpha_l}{2}\sigma F_{cd}(\Omega)\right] \\ &\quad -\frac{1}{2}\varepsilon_{sh}P_{sat}H_{sh}(\Omega)\sigma F_{sh} - \frac{1}{2}\varepsilon_{ch}P_{sat}H_{ch}(\Omega)\sigma F_{ch}\end{aligned} \tag{7.10}$$

$$\begin{aligned}F_{cd}(\Omega) &= \frac{1}{1-j\Omega\tau\alpha'}\left[\ln\left(\frac{1+GS(0)/P_{sat}-j\Omega\tau}{1+S(0)/P_{sat}-j\Omega\tau}\right)+\alpha'\ln\left(\frac{G_0}{G}\right)\right] \\ F_{sh} &= \ln\left(\frac{G_0}{G}\right) \\ F_{ch} &= -\frac{1}{\alpha'}\left[\frac{S(0)}{P_{sat}}(G-1)-\ln\left(\frac{G_0}{G}\right)\right]\end{aligned} \tag{7.11}$$

$$\begin{aligned}S(0) &= P_{sat}\left(\frac{1-\alpha'}{\alpha'}\right)\frac{1-(G/G_0)^{\alpha'}}{G-(G/G_0)^{\alpha'}} \\ \sigma &= \frac{P_0(0)}{P_0(0)+P_1(0)}\end{aligned} \tag{7.12}$$

In the above equations $\varepsilon_{sh}$ and $\varepsilon_{ch}$ are the strengths of the SHB and CH processes respectively. $H_{ch}$ and $H_{sh}$, the Fourier transforms of the nonlinear gain response due to CH and SHB respectively, are given by

$$H_{ch}(\Omega) = \frac{1}{(1 - j\Omega\tau_1)(1 - j\Omega\tau_2)}$$
$$H_{sh}(\Omega) = \frac{1}{1 - j\Omega\tau_2} \qquad (7.13)$$

Typical plots of $\rho$ and $\eta$ versus the absolute detuning frequency are given in Figs. 7.6 and 7.7. The parameters used in the calculations are: $\exp(\Gamma g_0 L) = 30$ dB, $\alpha' = 0.1$, $G = 18.6$ dB (obtained from 7.4), $\alpha_l = 4.0$, $P_{sat} = 10$ mW, $P_0 = 20$ µW, $P_1 = 200$ µW, $\tau = 0.1$ ns, $\tau_1 = 750$ fs, $\tau_2 = 150$ fs, $\varepsilon_{sh} = 10.0$ and $\varepsilon_{ch} = 2.5$. Fig. 7.6 shows that the $\rho$ versus $\Omega$ characteristic is asymmetric. Fig. 7.7 shows that $\eta$ is greater than unity for frequencies up to a few hundred GHz, i.e. frequency conversion with gain.

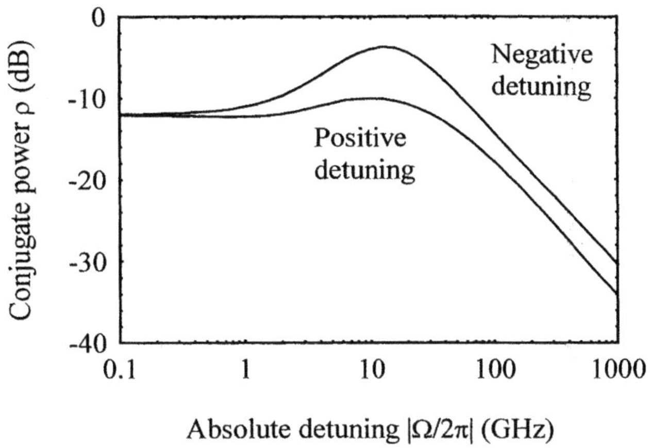


*Figure 7.6.* Normalised conjugate power versus absolute detuning frequency. The parameters used in the calculation are listed in the text (after [7] with permission of the IEEE).

## 7.1 SOA INTENSITY AND PHASE MODULATORS

SOAs can be used to intensity or phase modulate a CW optical signal. Intensity modulation can be achieved by simply modulating the amplifier current with a suitable data signal, as shown in Fig. 7.8 [8]. The current modulation modulates the amplifier gain and thereby the input optical signal. The modulation bandwidth is limited by the amplifier carrier lifetime. Much better performance is achievable using $LiNiO_3$ external modulators or electroabsorption modulators. The latter can also be integrated with a laser source. The linearity requirements of analog optical transmission systems usually preclude the possibility of using an SOA as an external modulator in such systems. This is due to carrier density induced second order nonlinear distortion. It has been shown that it is possible to use an SOA as an external modulator in a multichannel subcarrier multiplexed digital video link [9].

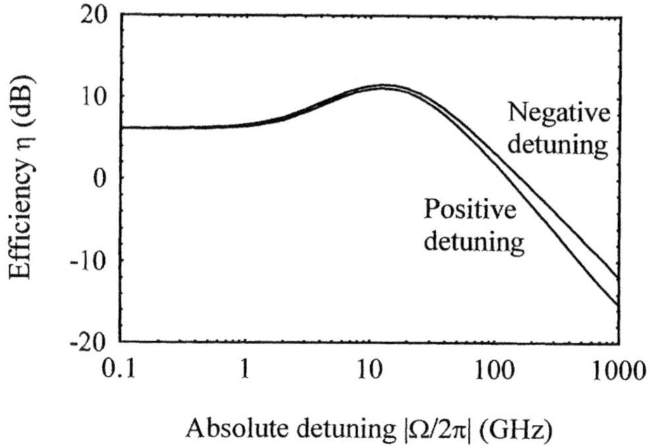


*Figure 7.7.* Conversion efficiency versus absolute detuning frequency. The parameters used are as for Fig. 7.6 (after [7] with permission of the IEEE).

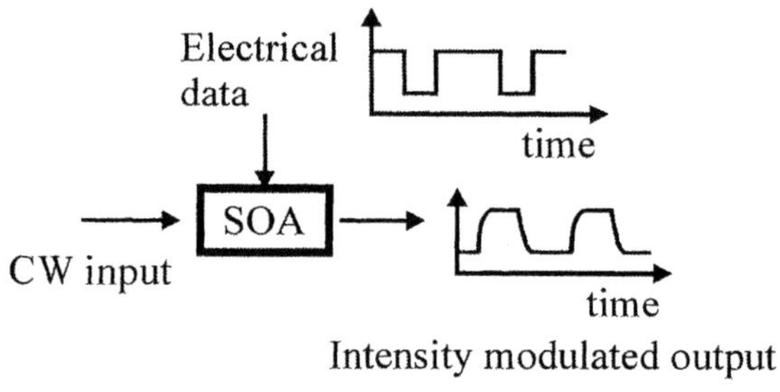


*Figure 7.8.* SOA used as an external intensity modulator. The output signal finite rise and fall times are due to the non-zero carrier lifetime

The phase of a signal propagating through an SOA can also be modulated via the carrier density dependence of the amplifier effective index. The phase modulation response is proportional to $\alpha_l$. As shown in Fig. 7.3, $\alpha_l$ increases as the signal photon energy nears the bandgap energy. This implies that the optimum phase response of the amplifier can be obtained by choosing an operating wavelength on the long-wavelength side of the amplifier gain spectrum. Phase modulators can be useful in coherent optical communication links.

In a single-section SOA phase modulator the ratio between the phase modulation index $\beta$ and amplitude or intensity modulation (IM) index $m$ is equal to $\alpha_l$. To avoid significant receiver penalty the ratio $\beta/m$ must be greater than 100. A two-section SOA, as shown in Fig. 7.9, can be used to suppress IM and so increase $\beta/m$ [10-11]. In this device the IM is cancelled by modulating the two electrodes in antiphase and by adjusting the modulation power to the front $P_f$ and back $P_b$ so that the two sections introduce equal amounts of IM. This gives a cancellation of the overall IM because the IM imposed by the two sections are $180^0$ out of phase. A similar reduction in the overall phase modulation can be avoided by using a non-uniform carrier density distribution, which leads to different $\alpha_l$ factors in the front and back sections of the device. The carrier density distribution is obtained by saturating the SOA and lowering the current in the back section.

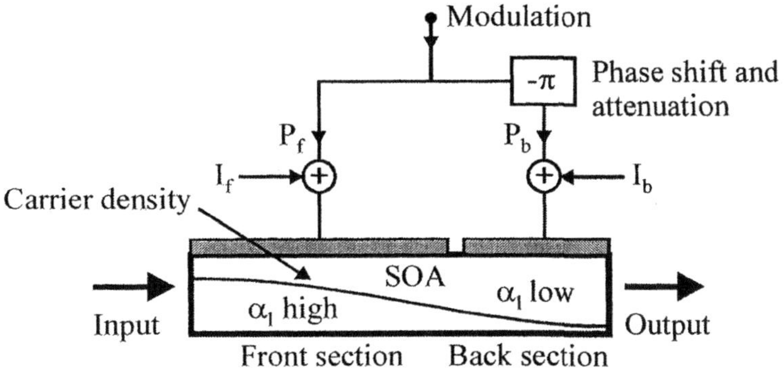


*Figure 7.9.* Two-section SOA used as a phase modulator with AM suppression (after [11] with permission of the IEE).

## 7.2 SOA WAVELENGTH CONVERTERS

All-optical wavelength converters are expected to play an important role in future broadband optical networks. Their most important function will be to avoid wavelength blocking in optical cross-connects in WDM networks. Wavelength converters will also increase the flexibility and capacity of a network using a fixed set of wavelengths. Wavelength conversion also can be used to centralise network management and allow easier protection switching [12-13]. In packet switching networks, tuneable wavelength converters can be used to resolve packet contention and reduce optical buffering requirements. A list of desirable properties of wavelength converters is given in Table 7.1.

*Table 7.1.* Ideal wavelength converter properties.

| |
|---|
| Data rate transparency |
| Low input power and high output power |
| Polarisation insensitive |
| Low noise figure |
| No extinction ratio degradation |
| Low chirp |
| Simple construction – integratable |

Wavelength converters can be based on structures exploiting the SOA nonlinearities. We first consider those structures exploiting XGM.

### 7.3.1 XGM wavelength converters

In Section 7.1.1 it was shown that it is possible to use XGM in an SOA to achieve wavelength conversion. There are two basic schemes used in XGM based wavelength converters; the co-propagating and counter-propagating schemes as shown in Fig. 7.10. In these schemes the pump and probe signals can be launched together or in opposite directions into the amplifier. The advantage of the counter-propagating scheme is that no optical filter is required to separate the modulated probe signal from the pump signal. However the co-propagating scheme has a superior noise performance [14]. In both schemes the converted signal (i.e. probe) is inverted compared to the input signal.

An expression for the small-signal response of the co-propagating scheme was derived in Section 7.1.1. This expression is in fact valid for all ratios of the input pump and probe powers provided the signals are co-propagating [15]. In the case of negligible waveguide losses, the conversion bandwidth is inversely proportional to the spontaneous carrier lifetime but enhanced by the amplifier gain. This bandwidth enhancement is due to propagation effects in the amplifier, which can be understood by reference to Fig. 7.11 [15]. The slow carrier response at the start of the amplifier leads to distortion of the optical pulses at the middle of the amplifier. The distorted pulses with strong leading edges then saturate the carriers in the last part of the device more rapidly leading to an enhanced bandwidth.

In the co-propagating scheme the conversion bandwidth can be further improved by cascading two or more SOAs separated by isolators (to reduce ASE induced gain saturation). The maximum number of XGM based wavelength converters that can be cascaded (typically between four to eight) is limited by pattern effects, ASE and jitter accumulation [16]. XGM wavelength converters also suffer from relatively large chirp because of the large gain modulation. The chirp increases the signal bandwidth and makes it more susceptible to fibre chromatic dispersion.

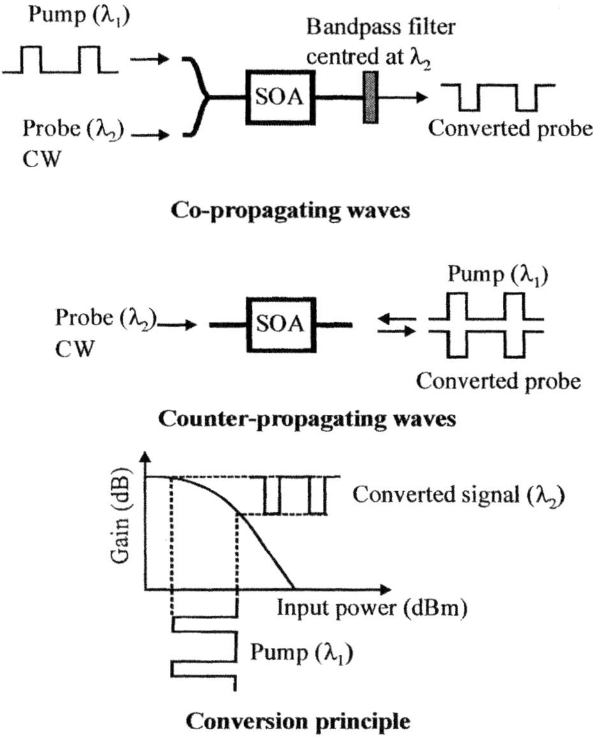


*Figure 7.10.* Co-propagating and counter-propagating wavelength converter configurations using XGM in an SOA. OBPF: Optical bandpass filter.

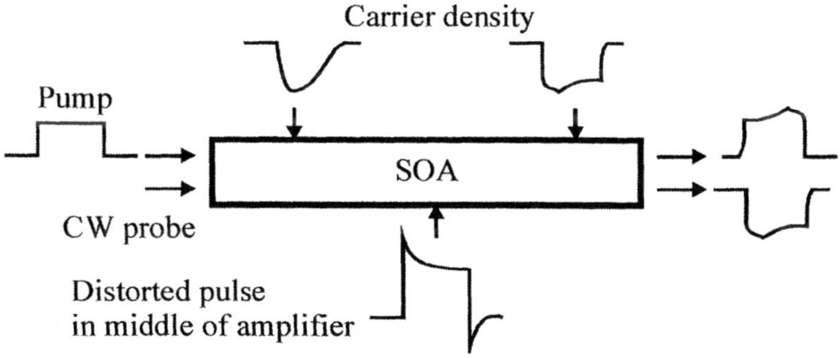


*Figure 7.11.* Propagation effects of XGM in an SOA (after [15] with permission of the IEE).

It is possible to compensate for the added chirp by passing the converted signal through an appropriate filter such as a fibre Bragg grating [17].

A number of other XGM based wavelength converter schemes have also been investigated. These include XGM of the amplifier ASE spectrum by an input pump signal [18]. The wavelength-converted output is obtained by selecting a slice of the amplifier output ASE spectrum centred at the desired wavelength. The converted signal will have a relatively wide optical bandwidth and consequently suffer from chromatic dispersion in fibre following the amplifier. However, this type of wavelength converter is simple, as it requires no CW probe source. It could find important applications in WDM access networks and LANs. The use of pump-induced birefringence in an SOA in addition to XGM can also be used to improve the converter extinction ratio and reduce power penalties [19]. In [20] wavelength conversion using XGM was demonstrated using a DFB laser monolithically integrated with an SOA, eliminating the need for an external laser and fibre coupler.

### 7.3.2 XPM wavelength converters

Wavelength converters can also be based on SOA XPM. These wavelength converters have a superior power efficiency compared to those devices based on XGM. They also have lower chirp because of the reduced gain modulation. In order to utilise XPM, one or more SOAs have to be placed in an interferometric configuration. The three basic types of interferometer used in SOA applications are described in Section 7.1.2. The most common type of interferometer used in XPM based wavelength converters is the MZI, two versions of which are shown in Fig. 7.12.

In the asymmetric MZI wavelength converter the CW input at $\lambda_2$ is split asymmetrically to each arm of the MZI by a coupler [21]. The intensity modulated signal at $\lambda_1$ saturates each SOA asymmetrically inducing different phase shifts in the input CW signal via carrier density induced refractive index changes. The output coupler recombines the split CW signals where they can interfere constructively or destructively. The actual state of interference depends on the relative phase difference between the interferometer arms, which relies both on the SOA bias currents and on the input optical powers. The symmetric MZI wavelength converter has the same operating principle. It requires an additional coupler but can use couplers with equal splitting ratios. Also the input signal is only fed to one of the SOAs.

If the powers of the CW signals in the upper and lower arms of the interferometer, just before interfering at the recombining coupler, are $P_u$ and $P_l$ respectively, then the power $P_o$ of the output signal is given by

$$P_o = P_u + P_l + 2\sqrt{P_u P_l}\cos\phi \tag{7.14}$$

where $\phi$ is the phase difference between the interfering waves. (7.14) is basically the interferometer transfer function. Whether the converted signal is inverted or non-inverted compared to the input data signal depends on the slope of the interferometer transfer function around the chosen operating point as shown in Fig. 7.13.

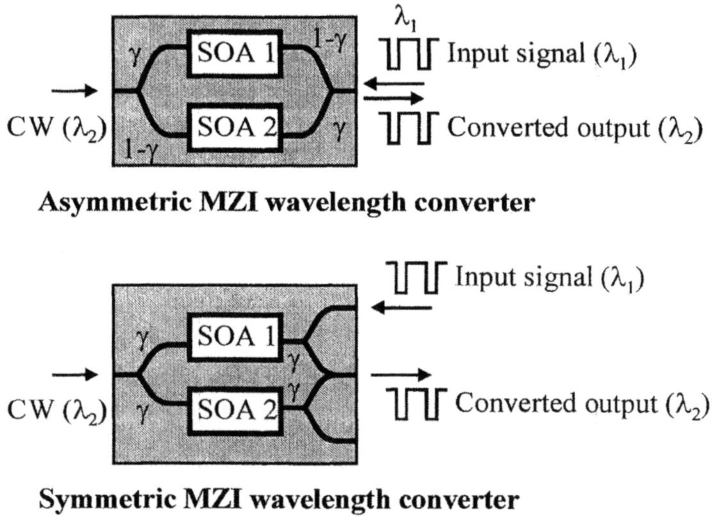


*Figure 7.12.* MZI wavelength converters based on XPM in SOAs. The coupler splitting ratios are different and equal in the asymmetric and symmetric converters respectively.

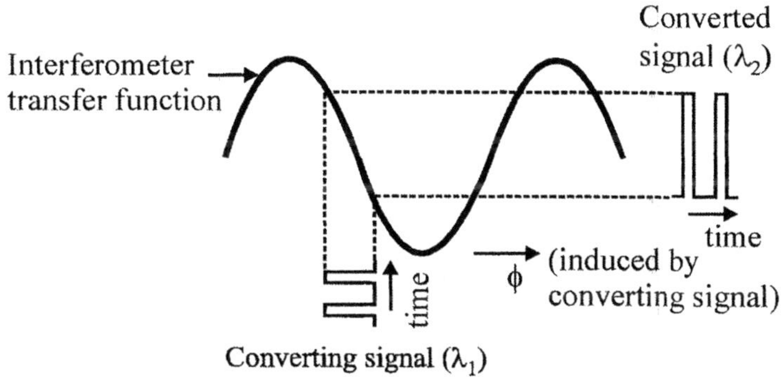


*Figure 7.13.* MZI wavelength converter transfer function.

In addition to phase changes the converting signal also leads to gain compression, which tends to unbalance the amplitude in the interferometer arms. This amplitude imbalance decreases the converter extinction ratio. It is possible to improve the conversion properties of the interferometer by the addition of a passive phase-shifter section in one of the interferometer arms. This enables pre-biasing of the interferometer in order to optimise the conversion efficiency [22].

It has been demonstrated that interferometric wavelength converters can be cascaded with very good performance [23]. Wavelength conversion can also be achieved without the use of an interferometer by using XPM in a single SOA followed by a fibre Bragg grating filter [24]. In this scheme the CW signal spectrum is chirped by XPM induced by the co-propagating pump signal. A non-inverted replica of the input data signal is obtained by selecting the red-shifted component of the CW spectrum using the filter.

Interferometric wavelength converters are amenable to optoelectronic integration. A PIC implementation of an MZI wavelength converter is shown in Fig. 7.14 [25]. The PIC contains a monolithically integrated DFB laser source and two inputs for symmetry and redundancy reasons each equipped with preamplifiers to compensate for variations in the input power.

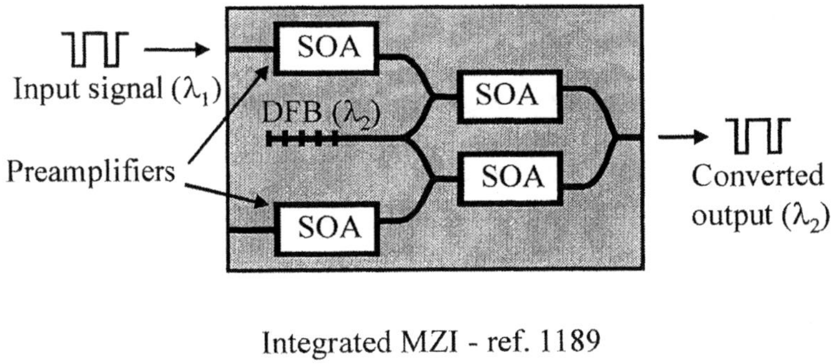


*Figure 7.14.* MZI wavelength converter with monolithically integrated DFB probe source (after [25] with permission of the IEEE)

### 7.3.3 FWM wavelength converters

SOA FWM can be used to construct wavelength converters [26-27]. The basic scheme is shown in Fig. 7.5, where CW pump and modulated probe input signals are injected into an SOA. The FWM process in the amplifier gives rise to a new conjugate signal that is a spectrally inverted replica of the input probe signal. An optical filter is then used to select the conjugate, i.e. wavelength-converted signal. FMW is transparent to both digital and analog modulated probe signals. However the conversion efficiency is relatively low for moderate values of frequency detuning between the pump and probe signals. For efficient FWM to occur in an SOA, the polarisation states of the pump and probe must be identical. In the simple wavelength converter of Fig. 5, this would necessitates polarisation control of either the pump or probe signal. This is because in a practical system, the polarisation state of the incoming pump (i.e. data) signal can vary slowly with time.

The co-polarised and orthogonal polarised dual-pump schemes, shown in Fig. 7.15, have the same structure but use different pump polarisations to achieve polarisation insensitive FWM [28]. In the former scheme the co-polarised pumps interact with the input data signal in an SOA to generate a new *non-conjugate* signal at the wavelength indicated in Fig. 7.15. In the latter scheme orthogonal polarised pumps interact with the input data signal as before to generate a new *conjugate* signal at the wavelength indicated in Fig. 7.15. In both schemes if the SOA is polarisation insensitive then the converted signal power will also be polarisation independent.

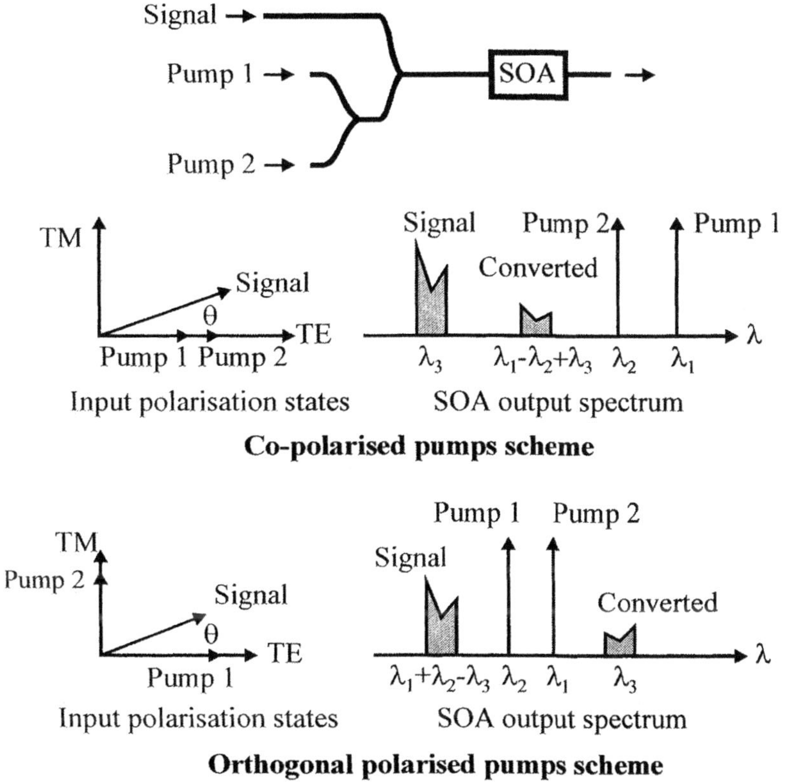


*Figure 7.15.* Co-polarised and orthogonal polarised dual-pump wavelength converters (after [28] with permission of the IEEE).

The polarisation diversity scheme, shown in Fig. 7.16, can also deliver polarisation insensitive FWM. In this scheme the input pump is polarised at 45° relative to the polarisation axes of polarisation beam splitter PBS1. This means that half of the pump power is delivered to each SOA along with a co-polarised component of the signal. These mix in each SOA to produce a conjugate signal with the same polarisation. The orthogonal polarised conjugate signals from the SOAs are recombined at the output in polarisation beamsplitter PBS2. If the SOAs have the same gain and conversion efficiencies then the scheme will be polarisation independent. A variant on the above uses bidirectional FWM for improved wavelength converter performance [29]. These schemes also improve the conversion efficiency over a greater detuning frequency band.

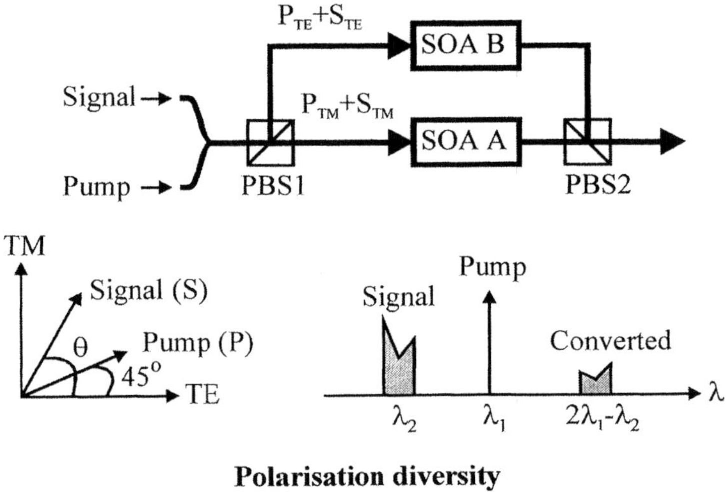


*Figure 7.16.* Polarisation-diversity FWM wavelength converter: PSB – polarisation beam splitter (after [28] with permission of the IEEE).

In the folded-path self-pumped configuration, shown in Fig. 7.17, wavelength conversion is achieved without the necessity of an external pump laser [30]. In this scheme the SOA lases at a wavelength determined by the fibre Bragg grating. The external data signal interacts with the lasing signal to produce a FWM converted signal. The high-reflection coating effectively doubles the interaction length between the input and lasing signals leading to greater conversion efficiency. The lasing action keeps the gain, and hence the nonlinearity, clamped at the threshold value.

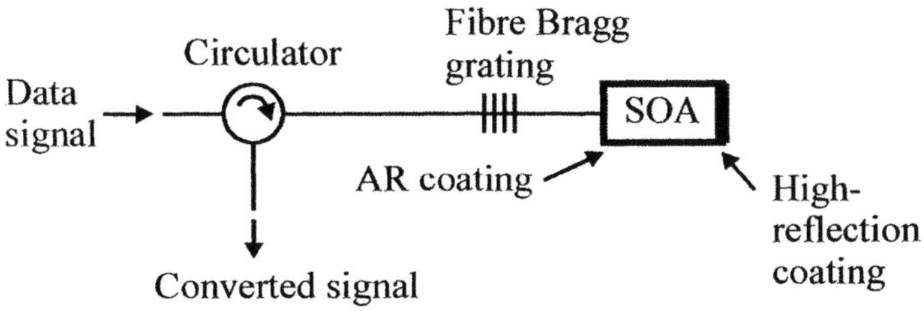


*Figure 7.17.* Folded-path self-pumped wavelength converter (after [30] with permission of the IEEE).

In the above schemes a tuneable narrow-band optical filter is required to separate the converted signal from the pump and input data signal. The design and practical implementation of such a filter can be difficult. The use of an optical filter can be avoided by using noncolinear FWM in a broad-area SOA as shown in Fig. 7.18 [31].

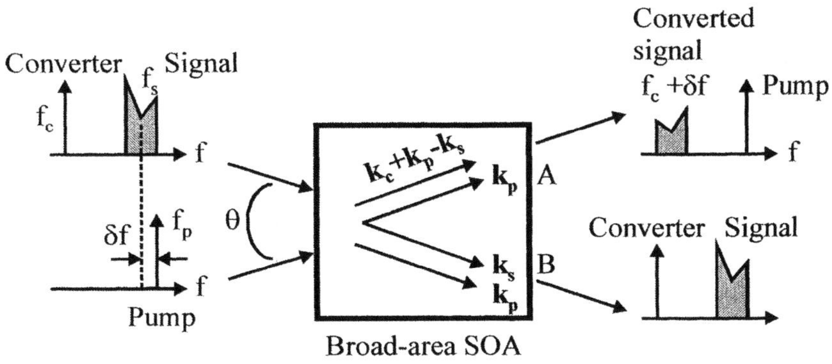


*Figure 7.18.* Filter-free FWM wavelength converter. Refractive index and gain gratings created in the SOA give rise to different propagation vector directions (**k**) for the various beams in the amplifier allowing spatial resolution of the converted signal from the converting signal at the amplifier output (after [31] with permission of the IEE).

This allows the various output beams from the amplifier to be spectrally and spatially resolved. Three input beams are used: Signal, converter and pump beams. The signal (frequency $f_s$) and converter ($f_c$) beams are combined together and injected into the SOA with angle $\theta/2$ relative to the amplifier axis. The pump ($f_p$) beam is injected into the amplifier with angle $\theta$ relative to the signal beam. The polarisation of the signal and pump beams are aligned in TE (or TM) polarisation, while the converter beam is aligned in TM (or TE) polarisation. The interference between the pump and signal beams creates refractive index and gain gratings in the amplifier. The converter beam is deflected by these gratings into a conjugate signal with frequency $f_c + f_p - f_s$. Because of phase matching considerations only one of the sidebands (the converted signal) is emitted from the SOA output port A in the same direction as the input signal beam. This spatial separation eliminates the need for a narrow band optical filter. The strong pump beam can be eliminated using an output polariser.

## 7.3 SOA OPTICAL GATES

Future high-speed WDM and TDM optical communication networks require high-speed optical switches (or gates) that can either be optically or electrically controlled [32-33]. Such optical switches will also form the basis for more advanced optical components such as add/drop multiplexers, demultiplexers and logic gates. The most desirable properties of an optical switch are: Fast switching speed, high on/off contrast ratio and the ability to be cascaded. Such optical switches can be constructed using SOAs.

There are many SOA gate configurations possible, the simplest of which consists of a single device as shown in Fig. 7.19 [34]. In this case the switch state is determined either by the electrical drive current to set the amplifier gain or by the optical pump power using XGM. The intrinsic switching time of the SOA is determined by the carrier recombination lifetime. The injection of a CW pump into the amplifier can be used to increase the switching speed by reducing the carrier lifetime via stimulated recombination [35]. It also can increase the number of gates that can be cascaded.

A variant on the optically controlled gate using counter-propagating waves in two SOAs is shown in Fig. 7.20 [36]. In the switch on state the control signal power is low so the gain of SOA 1 is high leading to a high output signal power. In the switch off state the control signal power is high causing SOA 1 to saturate. Simultaneously, because the data output power from SOA 1 is lower, the gain of SOA 2 increases leading to a higher control signal power. This means that a relatively small increase in the control input power causes a large change in the data output power so giving an improved switch contrast ratio.

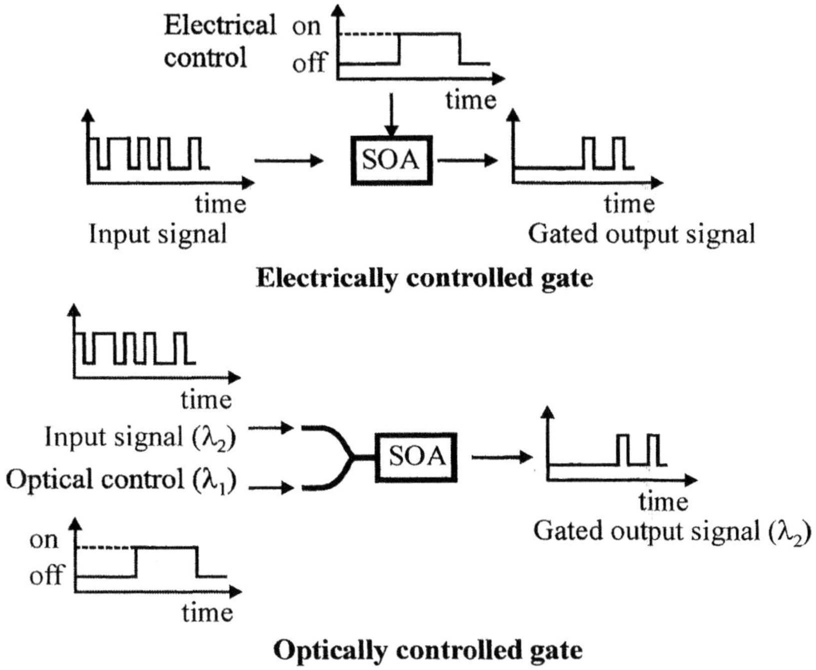


*Figure 7.19.* Basic electrically and optically controlled SOA gates.

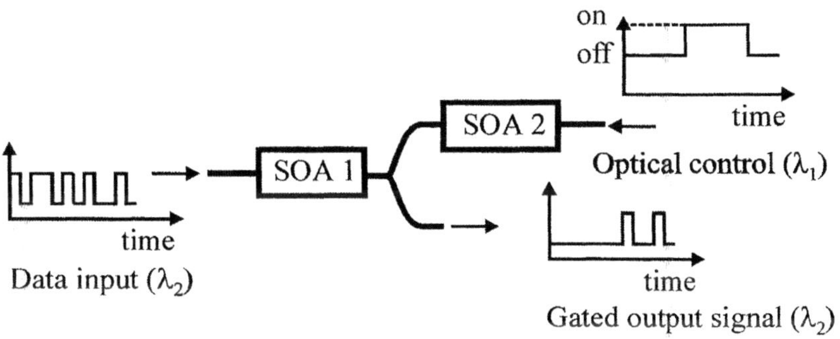


*Figure 7.20.* All-optical switch using counter-propagating operation in cascaded SOAs (after [36] with permission of the IEEE).

SOA gate arrays can be used to construct all-optical packet switching nodes where an incoming data packet can routed to a desired output port. The gate array can be fully integrated with input and output waveguides and couplers [37]. This technique helps to alleviate the mismatch the misalignment between the SOAs and the passive waveguides. However, it requires complicated multistep epitaxial growth and processing. An alternative solution is to use a hybrid structure using passive waveguides aligned to an SOA array as shown in Fig. 7.21 [38]. In the 2 x 2 switch module shown an incoming data packet can be routed to any output port by switching on the appropriate SOA.

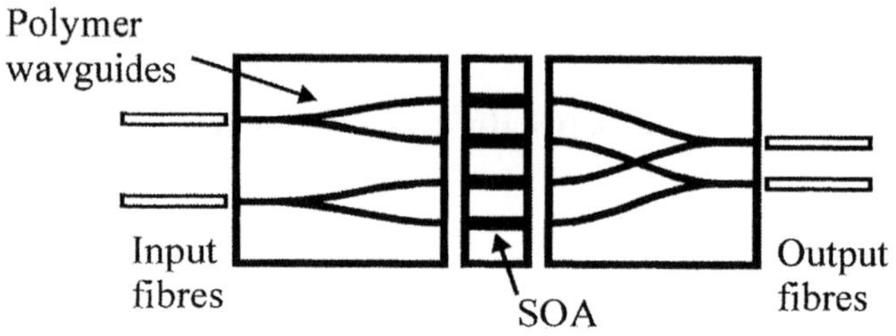


*Figure 7.21.* 2 x 2 hybrid SOA switch module (after [38] with permission of the IEEE).

SOA gate arrays can also be used to construct a wavelength selector as shown in Fig. 7.22 [39]. The device consists of two arrayed waveguide gratings (AWGs) and an eight-channel SOA array. Each AWG has eight

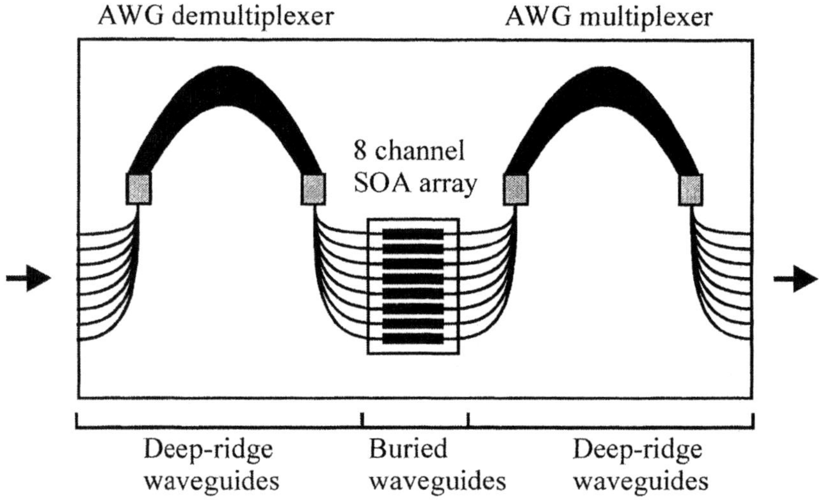


*Figure 7.22.* Mask layout of an integrated WDM channel selector (after [39] with permission of the IEEE).

input and eight output waveguides, two focussing slab regions and a phased array of 28 waveguides. The input WDM channels, with a channel spacing of 200 GHz, are fed into one of the input waveguides and demultiplexed by the first AWG. The desired signals are then selected by the SOA gate array and multiplexed into the output waveguide by the output AWG multiplexer.

A novel optical gate based on an SOA with an angled-grating Bragg is shown in Fig. 7.23 [40]. The grating has period $\Lambda$ with its planes orientated at angle $\theta_s$ with respect to the forward propagation direction $z$. The grating is embedded in a gain region pumped through a broad-area current stripe. An input beam incident on the grating is split into a transmitted beam and diffracted beam at the output.

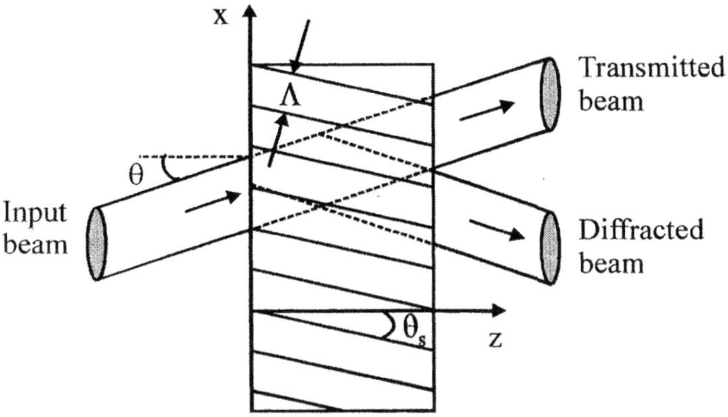


*Figure 7.23.* Angled-grating Bragg SOA geometry (after [40] with permission of the IEEE).

The diffraction efficiency is maximised when the input free-space wavelength $\lambda$ satisfies the Bragg condition

$$\lambda_B = 2n_r\Lambda\sin(\theta_B + \theta_S) \tag{7.15}$$

where $\theta_B$ is the Bragg angle and $n_r$ the amplifier refractive index. The diffraction efficiency is a very sensitive function of the detuning from the Bragg wavelength. The applied current via carrier-induced changes in the refractive index controls the diffraction efficiency. In theory a contrast ratio greater than 28 dB is possible with this device.

Interferometric configurations containing SOAs, similar to those used for wavelength converters, can also be used as optical switches. The most common interferometer used is the MZI. The basic principle is to use an optical control signal to change the carrier density in a SOA placed in one of the interferometer arms. This causes a refractive index change, and thereby a change in the phase of a passing data signal wave, which is used for switching. A symmetric MZI switch is shown in Fig. 7.24.

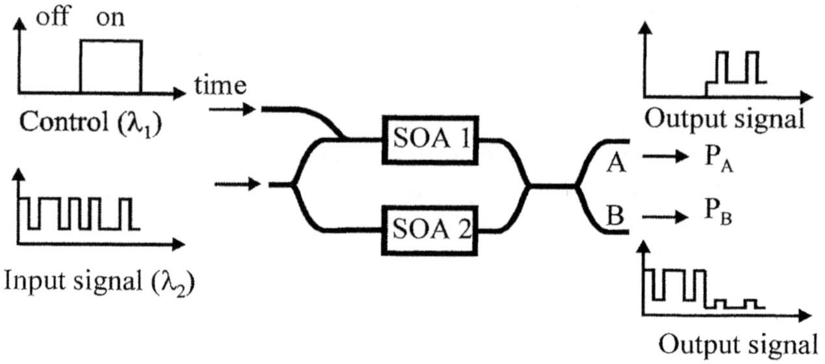


*Figure 7.24.* Optical switch based on a symmetric MZI configuration.

The amplifiers are equally biased to provide equal unsaturated gains. In the non-switched state, with no control signal, the input signal is directed to output port B with output power $P_X$ when constructive interference between the split signals is achieved. In the switched state, with the control signal high, the split signals destructively interfere causing the data signal to be directed to output port A. However, in this state the power in the split signals is not equal leading to some leakage of the data signal to port B and so a reduced contrast ratio. This problem can be overcome by the use of asymmetric MZI configurations [41]. The above switching mechanism cancels out the effect of the relatively slow carrier recombination time and makes ultrafast gating possible with switching times as short as 200 fs.

Interferometric optical switches are amenable to optoelectronic integration, an example of which is the hybrid-integrated symmetric MZI optical switch shown in Fig. 7.25. This consists of an SOA array mounted on a silica-based planar lightwave circuit (PLC) [42]. Compared to monolithic integration the passive silica waveguides have lower loss and lower group velocity dispersion and so are suitable for high switching speed.

Nonlinear optical loop mirrors (NOLMs), which are basically Sagnac interferometers, can also be used for high-speed optical switches. In the arrangement, shown in Fig. 7.26, switching is achieved by placing an SOA offset from the centre of an optical fibre loop mirror and injecting data into the loop via a 50:50 coupler [43].

The two counter-propagating data pulse streams arrive asynchronously at the SOA. A switching pulse is timed to arrive after one data pulse but just before its replica. The switching pulse power is adjusted to impart a phase change of $\pi$ radians onto the replica, so the data pulse is switched out when the two counter-propagating components interfere on their return to the coupler. This device is also known as a TOAD (terahertz optical asymmetric demultiplexer) because it can also be used to demultiplex high-speed TDM pulse streams.

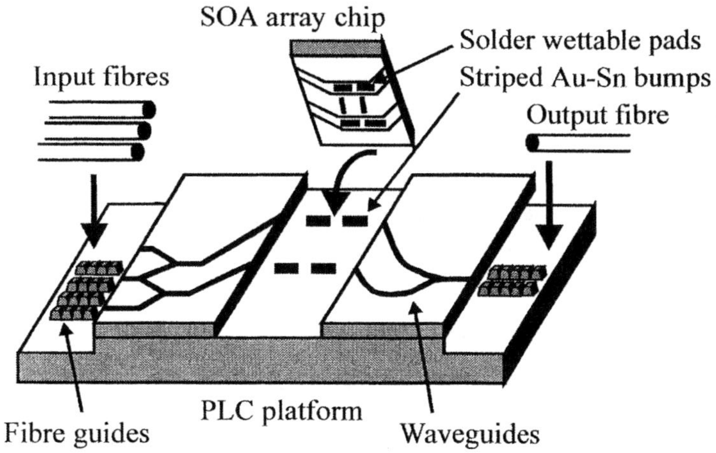


*Figure 7.25.* Hybrid-integrated symmetric MZI optical switch (after [42] with permission of the IEEE).

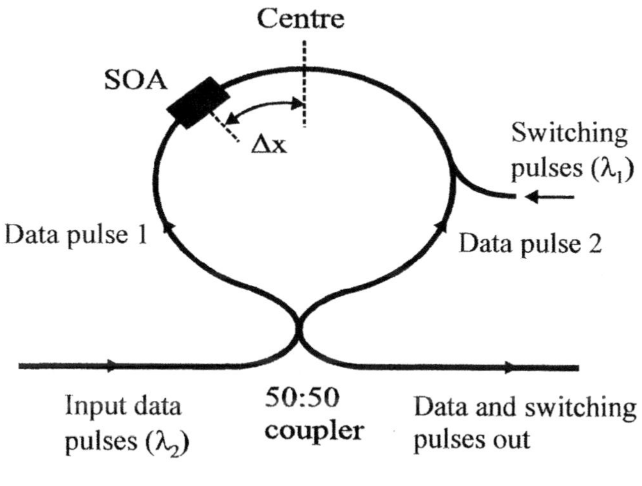


*Figure 7.26.* Optical switch using a TOAD.

The TOAD is quite bulky as it contains several fibre components. A more compact variant of the TOAD is the monolithically integrated all-SOA Sagnac interferometer device shown in Fig. 7.27 [44]. The device consists of four sections, the loop, a multimode interference waveguide amplifier

(MMIWA), the input port and output port. The device operation depends on the optical nonlinearity in the loop, as for the TOAD, but also on the nonlinear coupling effect in the MMIWA. When different currents are injected into electrodes $I_3$ and $I_4$ the asymmetrically distributed gain and gain saturation for counter-propagating signals in the loop make its operation similar to the TOAD for pulsed input signals. In the TOAD a conventional coupler is used to close the loop. In this device the MMIWA provides nonlinear coupling. The lateral distribution of optical intensity in the MMIWA and hence the power distribution of the counter-propagating signals coupled into the loop depend on the input power level. The MMIWA can then enhance the switching performance.

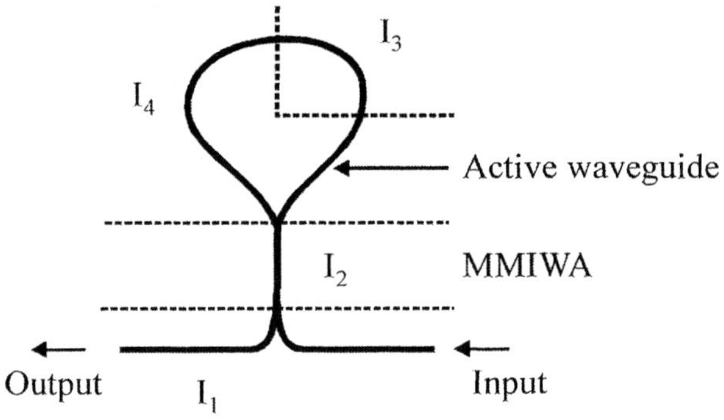


*Figure 7.27.* Layout of the all-SOA Sagnac interferometer. The device has a loop radius of 300 μm and an MMIWA length of 480 μm (after [44] with permission of the IEEE).

An optical switch can also be constructed from two crossing single-mode waveguides with an intersection region as shown in Fig. 7.28 [45]. The device is fabricated from InP/InGaAsP material. This device is also known as a total internal reflection switch. The intersection region is twice as wide as the waveguides and maintains two modes. Current can be applied to half of the intersection region, i.e. the amplifier section. In the absence of current the intersection region acts as a two-mode interference section and the ratio of light intensity in the output waveguides depends on the difference in effective index for the $0^{th}$ and $1^{st}$ order modes and the length of the section. The guide effective refractive index can be lowered in half of the intersection region by applying a current. Hence optical switching can be achieved through current control of the waveguide effective index.

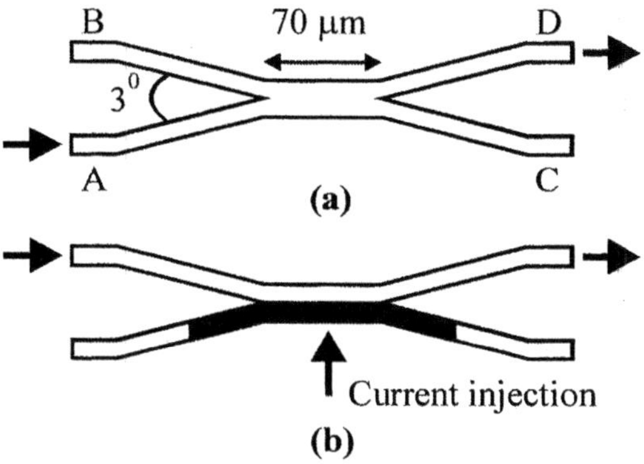


*Figure 7.28.* Schematic diagram of the total internal reflection switch, showing the two crossing waveguides and the optical amplifier. In (a) with no current injected, light goes from input port A to output port D. In (b) the refractive index is lowered so light goes from input port B to port D (after [45] with permission of the IEEE).

## 7.4 SOA LOGIC

Optical logic can be useful for all-optical signal processing applications in high-speed optical networks. In this section we consider some SOA based configurations for realising basic all-optical logic, three of which are shown in Fig. 7.29.

The novel XOR gate of Fig. 7.29(a) is based on XPOM in the SOA between two input data signals of equal power, equal polarisation and similar wavelength [46]. A compensator and polariser are placed after the SOA and tuned to block light passing through. When both inputs are on the output is off. When both inputs are off the output is off. However, when either of the inputs are off the output signal polarisation is different to what it was when both inputs were high. In this case the polariser does not block the output signal entirely and the output is on. This behaviour corresponds to logic XOR. It is also possible to obtain XOR operation using FWM in SOAs [47] and also in SOA fibre Sagnac interferometers [48].

The OR gate of Fig. 7.29(b) is based on a Michelson interferometer with two data signals coupled into port 1 and CW light coupled into port 2 [49]. The total input signal power controls the phase shift experienced by the CW input as it makes a round-trip from port 2 to the reflective end of the lower arm of the interferometer and back to port 2, where interference occurs between the returning signal and the CW input. The MI is designed such that destructive interference occurs when there is no input signal power and constructive interference when signal power is present. The power launched into the upper arm of the MI depends on the sum of the power of the two input signals so the power level is greater for two marks than for a mark and a space. This difference can be suppressed using the interferometer sinusoidal transfer function where the OR output is effectively clamped for large input signal powers (see Fig. 7.13).

The simple NOR gate of Fig. 7.29(c) is based on XGM in an SOA between two input data signals and an input CW probe [50]. When there are no input signals the output probe signal is high as there is no gain saturation. When either of the input signals is high the amplifier saturates causing the output probe to be low. This corresponds to NOR operation.

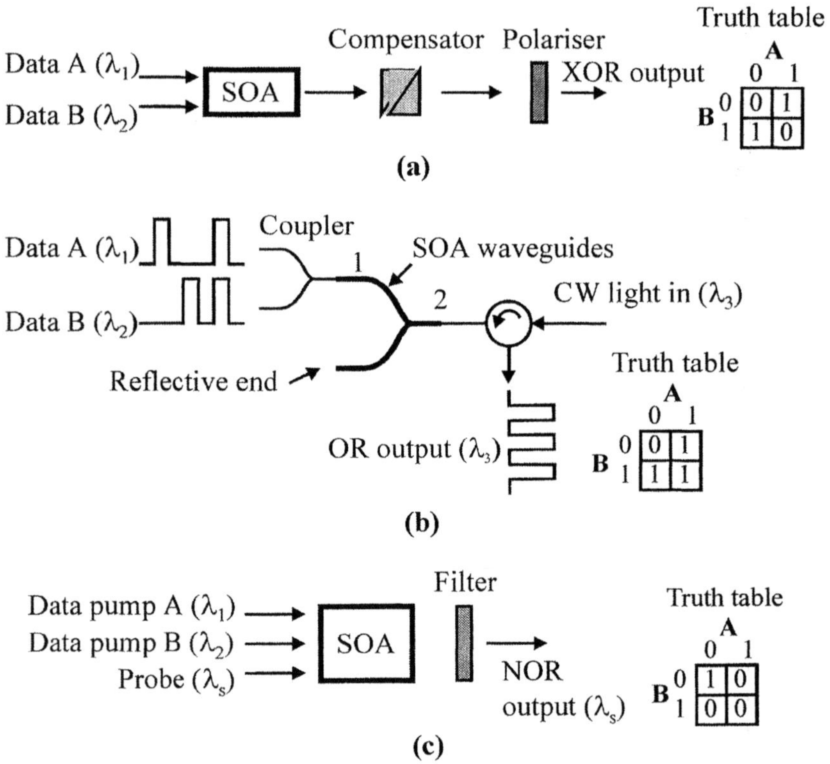


*Figure 7.29.* SOA logic gates, (a) XOR gate, (b) OR gate, (c) NOR gate (after [46,49,50] with permission of the IEEE and IEE).

## 7.6 SOA ADD/DROP MULTIPLEXERS

Optical time division demultiplexers (OTDDMs) and add/drop multiplexers (ADMs) are key components required by optical time division multiplexed (OTDM) network nodes. In an ADM one channel is dropped from an incoming TDM data stream leaving the other channels undisturbed. A new channel can be added by inserting data pulses into the vacant time slot. There are a number of techniques possible for realising OTDDMs and ADMs using SOAs These include FWM, MZI configurations and TOADs. In WDM networks there is also a requirement for add/drop wavelength multiplexers (ADWMs).

#### OTDM DEMULTIPLEXING USING FWM

An OTDM demultiplexer can be realised using FWM in an SOA [51] as shown in Fig. 7.30. In this technique a chirped clock pulse train probe is injected along with a modulated signal pulse stream into an SOA. Each individual signal pulse interacts with a different wavelength component of the clock in a FWM process. The polarisation controllers are used to match the clock and signal polarisations in order to maximise the FWM efficiency. Therefore, FWM signals generated by the input signals in different time slots have different wavelengths. The FWM signals can then be separated using a WDM demultiplexer.

#### OTDM MZI ADD/DROP MULTIPLEXER

MZI switches incorporating SOAs can also be used as ADMs. Many configurations are possible, one of which is shown in Fig. 7.31 [52]. In this configuration the input data signal at 40 Gb/s is split into two drive signals. One of the drive signals is delayed by a half a bit period (i.e. a half a single time slot). The interferometer is configured such that when an undelayed signal pulse is present in the upper arm of the interferometer an input 10 GHz pulse is directed to the drop port. At the same time the 3x10 GHz pulse stream is directed to the through port. When the delayed signal pulse is present in the lower arm of the interferometer the data is directed away from the drop port. The amplitudes of the drop and through pulses are modified by the SOA gain saturation induced by the input data pulses so pulse amplification and reshaping also occurs, i.e. the device functions as a 2R regenerator. If it is combined with optical clock recovery for retiming it will function as a 3R regenerator. Data can be added to the vacant time slot in the output data simply by sending the add channel data pulses to the add port.

A TOAD using the conventional structure of Fig. 7.26 or MZI structures can also be used as OTDM demultiplexers [53-54].

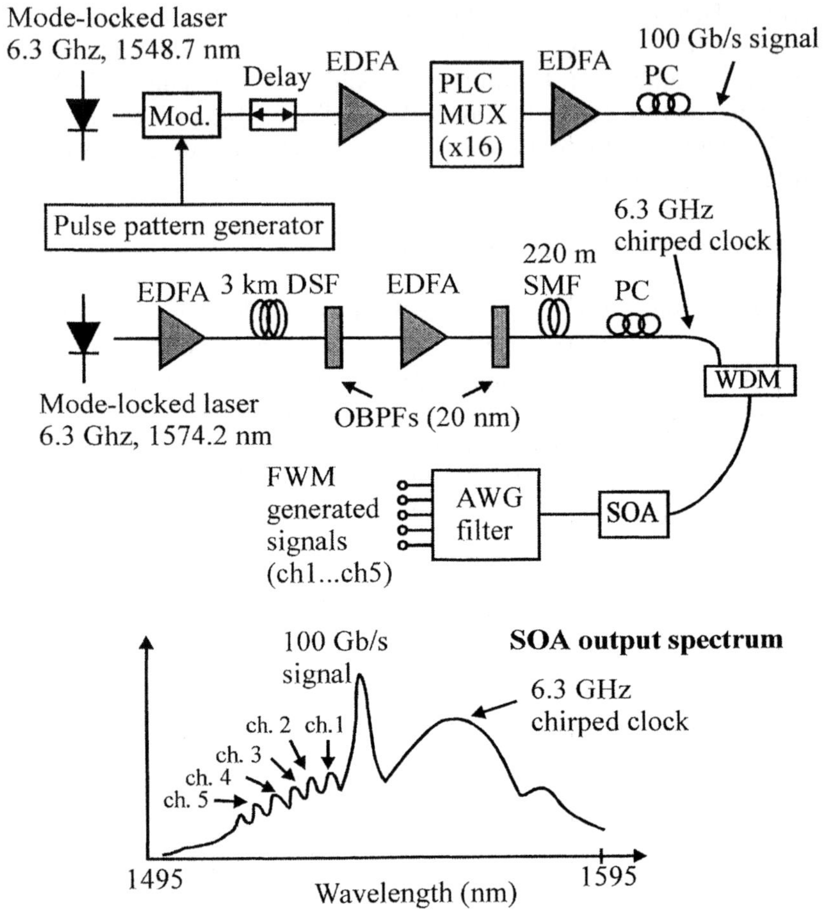


*Figure 7.30.* OTDM demultiplexer using FWM in an SOA: PC: polaisation controller, DSF: dispersion shifted fibre, SMF: single-mode fibre, AWG: arrayed waveguide grating. A 100 Gb/s signal is generated by optical time division multiplexing a 6.3 GHz signal obtained from a mode-locked laser 16 times (after [51] with permission of the IEEE).

## WDM ADD/DROP MULTIPLEXER

The ability to add and drop wavelength channels in WDM networks is useful for wavelength routing. The function of a wavelength ADM is to separate a particular wavelength channel without interference from adjacent channels. This can be achieved by a wavelength demultiplexer such as an AWG or by using an integrated SOA with a tuneable filter as shown in Fig. 7.32 [55]. The tuneable filter transfer function has a spectral width of 0.9 nm around the selected wavelength. The filter can be tuned by changing its injected current. The selected wavelength channel is reflected by the filter, amplified a second time by the MQW section and extracted to a drop port using a circulator. The remaining channels pass through the filter section to which it is a simple matter to add a new wavelength channel.

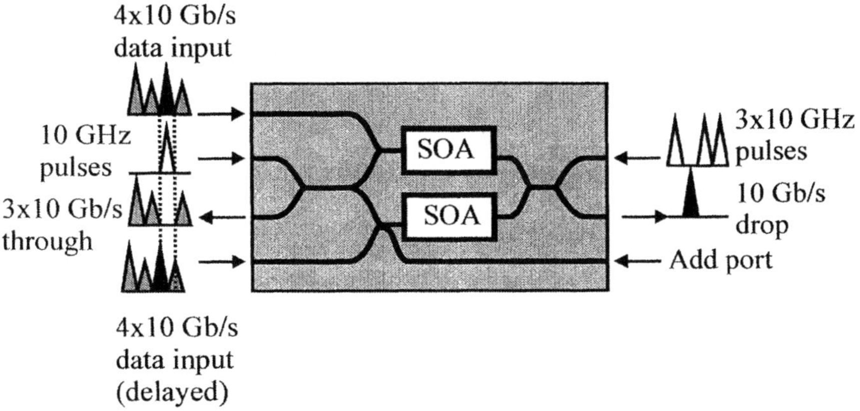


*Figure 7.31.* MZI ADM (after [52] with permission of the IEEE).

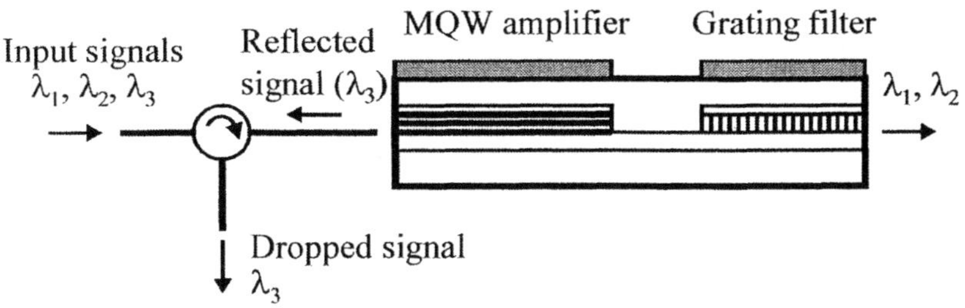


*Figure 7.32.* Tuneable SOA-filter wavelength ADM (after [55] with permission of the IEEE).

## 7.7 SOA PULSE GENERATORS

High repetition-rate wavelength tuneable pulses are required in high-speed OTDM WDM communication links. At high frequencies (> 10 GHz) it is difficult and expensive to generate such pulses by electronic means. One optical technique is to use a mode-locked fibre ring laser incorporating an SOA as shown in Fig. 7.33 [56]. The simultaneous generation of four tuneable wavelength channels each transmitting 12 ps pulses at 2.5 GHz has been demonstrated using a mode locked semiconductor laser based on an SOA and a diffraction grating [57].

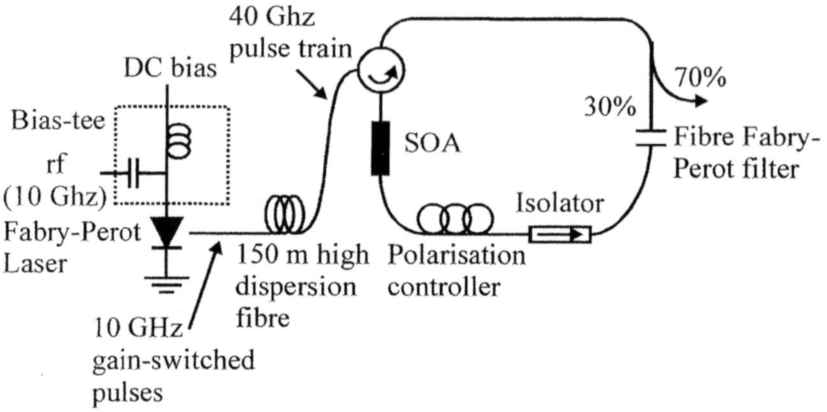


*Figure 7.33.* Optical pulse generation using a mode-locked fibre ring laser incorporating an SOA (after [56] with permission of the IEEE).

In Fig. 7.33 the Fabry-Perot laser is gain-switched by sinusoidal modulation at 10 GHz. After transmission through 150 m of high dispersion fibre each spectral mode in the gain-switched pulses is delayed by 25 ps with respect to its nearest mode. So each dispersed pulse creates a sequence of pulses of different wavelengths separated by 25 ps, giving an effective repetition rate of 40 GHz. However, the pulses vary in wavelength corresponding to the Fabry-Perot modes of the laser. This can give rise to significant dispersion when the pulses are transmitted down a fibre link. The pulses can be converted to pulses of the same (tuneable) wavelength by using a fibre ring laser with an SOA as the active element. A laser can oscillate on many longitudinal modes. These modes can be coupled together by external means, which locks their phases to each other. The modes can then be regarded as the components of a Fourier series expansion of a periodic function (in this case a pulse train) of fundamental frequency $c/2n_rL$ where $n_r$ is the refractive index of the laser and $L$ its length. In Fig. 7.33 the ring length is 15.7 m giving a fundamental frequency $f_c$ of 12.31 MHz. The 40 GHz optical pulse stream is inserted into the ring by a circulator, causing the gain of the SOA to be optically modulated. This mode-locks the laser. By choosing an appropriate modulation frequency of the laser, 10.188 GHz, it is possible to mode lock the laser at 40.752 GHz (the 3310$^{th}$ harmonic of $f_c$). The output wavelength of the fibre laser is selected using a fibre Fabry-Perot filter. The output power is coupled from the fibre ring using a fibre coupler.

## 7.8 SOA CLOCK RECOVERY

In OTDM systems, clock recovery is required in optical receivers and in 3R regenerators. At high speeds clock recovery is best achieved using an optical solution. Non-SOA based techniques include injection locked lasers and self-pulsating two-section DFB lasers. SOA based clock recovery techniques include using a phase comparator based on SOA FWM [58], SOA self-phase modulation [59], TOADs and symmetrical MZIs. All optical clock division can be achieved using a SOA loop mirror [60].

Another technique, shown in Fig. 7.34 uses a phase locked loop with an SOA based interferometric switch [61]. In this configuration the OTDM data

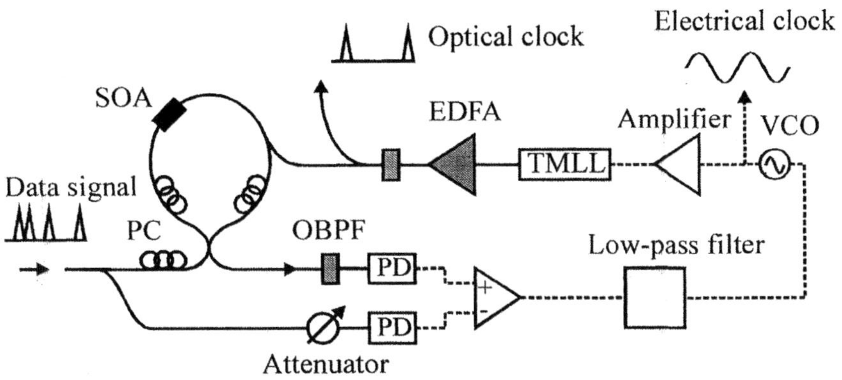


*Figure 7.34.* Optical clock recovery using an opto-electronic phase locked loop and interferometric SOA switch, PD: photodiode, TMLL: tuneable mode locked laser, OBPF: optical bandpass filter, PC: polarisation controller, VCO: voltage controlled oscillator (after [61] with permission of the IEEE).

signal is coupled to the SOA loop mirror, which is driven by an optical control pulse train generated by a tuneable mode locked laser (TMLL), whose repetition frequency is determined by a voltage-controlled oscillator (VCO). The output signal from the loop mirror is detected by a slow photodiode. A fraction of the input signal is switched from the loop mirror at the repetition rate of the control pulses. When the VCO frequency is equal to the base frequency of the input signal, the switched components of the input signal have constant phase within a time slot. In this case the output signal from the photodiode becomes a DC signal whose amplitude is proportional to the phase difference between the input signal pulses and control pulse train, i.e. the optical switch acts as a phase comparator. However this error signal only has a single polarity so there is no discrimination between negative and positive phase differences. This problem can be overcome by detecting the signal using a second slow photodiode. The output signal from this photodiode is subtracted from the error signal. The resulting signal is sent to the VCO via a low-pass filter. This closes the loop and locks the VCO frequency to the base frequency of the input data signal. The optical clock pulses can then be extracted from the output of the TMLL using a coupler.

## 7.9 SOA DISPERSION COMPENSATORS

In optical communication networks using conventional non-dispersion shifted fibre, chromatic dispersion is a major limiting factor on the transmission distance achievable at high data rates. All-optical dispersion compensation can be achieved using XPM in an SOA followed by a chirped fibre Bragg grating [17] and also using FWM in a polarisation diversity scheme similar to Fig. 7.16 [62]. In [63] three SOAs were used to improve the FWM efficiency of an SOA dispersion compensator. In [64] dispersion compensation was achieved using FWM in an SOA integrated with a DFB pump laser.

## 7.10 SOA DETECTORS

An SOA can be used to detect an optical signal [65-66]. This allows the amplifier to be used to perform a variety of functions such as simultaneous tapping and amplification of data, monitoring of input power and amplifier stabilisation using a feedback loop. This detection property arises from the interaction of the amplified light with the amplifier gain medium. When the amplifier is operated below transparency light photons are absorbed giving rise to electron-hole pairs. The resulting detected current can be detected at the amplifier electrode and converted to a voltage by passing it through a load resistor as shown in Fig. 7.35. In this case the detected current increases with an increase in the light power. When the amplifier is operated above transparency, the injected carriers from the bias current are depleted leading to a decrease in the total current at the amplifier electrode. In this case the detected current becomes more negative as the light power increases. Because of these two different mechanisms, the detected current will change polarity at transparency. The figure of merit of an SOA detector is the responsivity $R_d$ defined by

$$R_d = \frac{v_L(\omega)}{p(\omega)} \quad mV/mW \tag{7.16}$$

where $p(\omega)$ is a small-signal sinusoidal input optical power at angular frequency $\omega$ and $v_L(\omega)$ the resulting small-signal voltage dropped across a load resistor. A typical responsivity curve is shown in Fig. 7.35. As the detection mechanism is based on signal induced carrier recombination, the detection bandwidth is given by $B = 1/(2\pi\tau)$ where $\tau$ is the differential carrier lifetime. This means in practice that the maximum bandwidth achievable is of the order of a few GHz. The responsivity can be improved by using a two-section SOA [67].

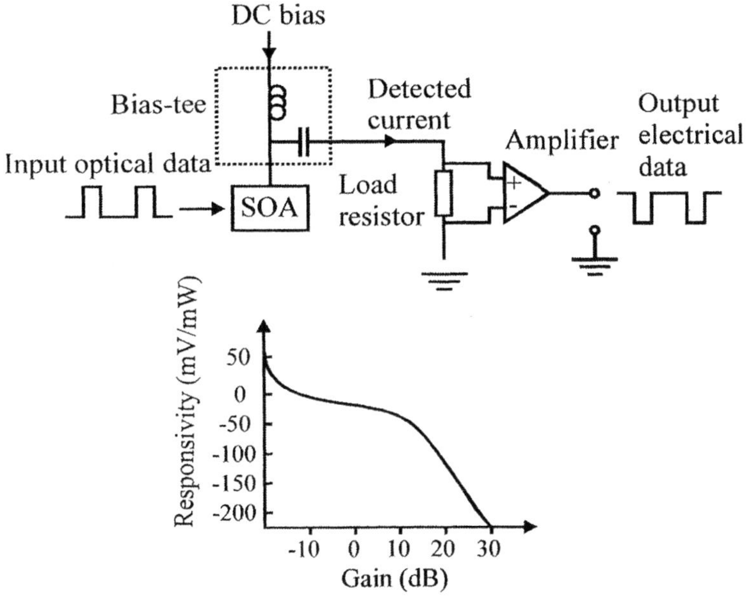


*Figure 7.35.* SOA detector and typical responsivity versus the amplifier gain characteristic. The load resistance is 50 Ω.

## REFERENCES


1. A. Mecozzi, Small-signal theory of wavelength converters based on cross-gain modulation in semiconductor optical amplifiers, *IEEE Photon. Tech. Lett.*, **8**, 1471 –1473 (1996).
2. D.A.O. Davies, Small-signal analysis of wavelength conversion in semiconductor laser amplifiers via gain saturation, *IEEE Photon. Tech. Lett.*, **7**, 617-619 (1995).
3. M. Osinski and J. Buus, Linewidth broadening factor in semiconductor lasers - An overview, *J. Lightwave Technol*, **23**, 9-28 (1987).
4. B.E. Saleh and M.C. Teich, *Fundamentals of Photonics*, John Wiley: New York (1991).
5. H. Soto, D. Erasme and G. Guekos, Cross-polarization modulation in semiconductor optical amplifiers, *IEEE Photon. Technol. Lett.*, **11**, 970-972 (1999).

6. J. Agrawal, Population pulsations and nondegenerate four-wave mixing in semiconductor lasers and amplifiers, *J. Opt. Soc. Am. B*, **5**, 147-159 (1988).
7. A. Mecozzi, S. Scotti, A. D'Ottavi, E. Iannone and P. Spano, Four-wave mixing in traveling-wave semiconductor amplifiers, *IEEE J. Quantum Electron.*, **3**, 689-699 (1995).
8. M.A. Ali, A.F. Elrefaie and S.A. Ahmed, Simulation of 12.5 Gb/s lightwave optical time-division multiplexer using semiconductor optical amplifiers as external modulators, *IEEE Photon. Technol. Lett.*, **4**, 280-383 (1992).
9. M.A. Ali and G. Metivier, Performance analysis of multichannel 16/64-QAM CATV distribution network using semiconductor optical amplifier as an external modulator, *IEEE Photon. Technol. Lett.*, **9**, 690-692 (1997).
10. M. Le Ligne, V. Le Gleau, D.H.Bon Hoa and M.J. Connelly, Two-section semiconductor laser amplifier used as a phase modulator in a DPSK heterodyne systems, *J. Opt. Communications*, 15, 82-85 (1994).
11. T.N. Nielsen, N. Storkfelt, U. Gliese, B. Mikkelsen, T. Durhuus, K.E. Stubkjaer, B. Fernier, F. Leblond, and A. Accard, Cancellation of inherent AM in semiconductor optical amplifier phase modulators, *Electron. Lett.*, **28**, 235-236 (1992).
12. T. Durhuus, B. Mikkelsen, C. Joergensen, S. Lykke Danielsen and K.E. Stubkjaer, All-optical wavelength conversion by semiconductor optical amplifiers,, *J. Lightwave Technol.*, 14, 942-954 (1996).
13. E. Iannone, R. Sabella, L. De Stefano and F. Valeri, All-optical wavelength conversion in optical multicarrier networks, *IEEE Trans. on Communications*, **44**, 716-724 (1996).
14. K. Obermann, S. Kindt, D. Breuer, K. Petermann, C. Schmidt, S. Diez and H.G. Weber, Noise characteristics of semiconductor-optical amplifiers used for wavelength conversion via cross-gain and cross-phase modulation, *IEEE Photon. Technol. Lett.*, **9**, 312-314 (1997).
15. D.D. Marcenac, A.E. Kelly, D. Nesset and D.A.O. Davies, Bandwidth enhancement of wavelength conversion via cross-gain modulation by semiconductor optical amplifier cascade, *Electron. Lett.*, **31**, 1442-1443 (1995).
16. X. Zheng, F. Liu and A. Kloch, Experimental investigation of the cascadability of a cross-gain modulation wavelength converter, *IEEE Photon. Technol. Lett.*, **12**, 272-274 (2000).
17. H. Yu, D. Mahgerefteh, P.S. Cho and J. Goldhar, Improved transmission of chirped signals from semiconductor optical devices by pulse reshaping using a fiber Bragg grating filter, *J. Lightwave Technol.*, **17**, 898-903 (1999).
18. L. Deming, N.J. Hong and L. Chao, Wavelength conversion based on cross-gain modulation of ASE spectrum of SOA, *IEEE Photon. Technol. Lett*, **12**, 1222-1224 (2000).
19. H.J. Lee, M. Sohn, K. Kim and H.G. Kim, Wavelength dependent performance of a wavelength converter based on cross-gain modulation and birefringence of a semiconductor optical amplifier, *IEEE Photon. Technol. Lett.*, 11, 185-187 (1999).
20. M.F.C. Stephens, D. Nesset, K.A. Williams, A.E. Kelly, R.V. Penty, I.H. White, and M.J. Fice, Wavelength conversion at 40 Gbit/s via cross-gain modulation in distributed feedback laser integrated with semiconductor optical amplifier, *Electron. Lett.*, **35**, 1762-1764 (1999).
21. F. Ratovelomanana, N. Vodjdani, A. Enard, G. Glastre, D. Rondi, R. Blondeau, C. Joergenesen, T. Durhuus, B. Mikkelsen, K.E. Stubkjaer, A. Jourdan and G. Soulage, An all-optical wavelength-converter with semiconductor optical amplifiers monolithically integrated in an asymmetric passive Mach-Zehnder interferometer, *IEEE Photon. Technol. Lett.*, **7**, 992-994 (1995).
22. F. Ratovelomanana, N. Vodjdani, A. Enard, G. Glastre, D. Rondi, R. Blondeau, A. Dupas, L. Billes and J.C. Simon, Regeneration improvement in all-optical wavelength converter,

based on a Mach-Zehnder interferometer by means of phase-shifter section, *Electron. Lett.*, 33, 1629-1630 (1997).
23. R.J.S. Pedersen, M. Nissov, B. Mikkelsen, H.N. Poulsen, K.E. Stubkjaer, M. Gustavsson, W. van Berlo and M. Janson, Transmission through a cascade of 10 all-optical interferometric wavelength converter spans at 10 Gbit/s, Pedersen, *Electron.. Lett.*, **32**, 1034-1035 (1996).
24. P.S. Cho, D. Mahgerefteh, J. Goldhar, L.G. Joneckis and G.L. Burdge, High performance non-interferometric semiconductor optical amplifier/fibre Bragg grating wavelength converter, *IEE Electron. Lett.*, **34**, 371-373 (1998).
25. L.H. Spiekman, U. Koren, M.D. Chien, B.I. Miller, J.M. Wiesenfeld and J.S. Perino, All-optical Mach-Zehnder wavelength converter with monolithically integrated DFB probe source, *IEEE Photon. Technol. Lett.*, **9**, 1349-1351 (1997).
26. S. Diez, C. Schmidt, R. Ludwig, H.G. Weber, K. Obermann, S. Kindt, I. Koltchanov, and K. Petermann, Four-wave mixing in semiconductor optical amplifiers for frequency conversion and fast optical switching, *IEEE J. Sel. Topics in Quantum Electron.*, **3**, 1131-1145 (1997).
27. D.F. Geraghty, R.B. Lee, M. Verdiell, M. Ziari , A. Mathur and K.J. Vahala, Wavelength conversion for WDM communication systems using four-wave mixing in semiconductor optical amplifiers, *IEEE J. Sel. Topics in Quantum Electron.*, **3**, 1146-1155 (1997).
28. J.P.R. Lacey, M.A. Summerfield and S.J. Madden, Tunability of polarization-insensitive wavelength converters based on four-wave mixing in semiconductor optical amplifiers, *J. Lightwave Technol.* **16**, 2419-2427 (1998).
29. G. Contestabile, A. D'Ottavi, E. Martelli, A. Mecozzi and P. Spano, Polarization- and interval-independent wavelength conversion at 2.5 Gb/s by means of bidirectional four-wave mixing in semiconductor optical amplifiers, *IEEE Photon. Technol. Lett.*, **12**, 852-854 (2000).
30. G. Hunziker, R. Paiella, M. Ziari, A. Mathur and K.J. Vahala, Folded-path self-pumped wavelength converter based on four-wave mixing in a semiconductor optical amplifier, *IEEE Photon. Technol. Lett.*, **9**, 1352-1354 (1997).
31. D.X. Zhu, D. Tishinin, K. Uppal, S. Dubovitsky, J. Burger, W.H. Steier and D.P. Dapkus, Filter-free four-wave mixing wavelength conversion in semiconductor optical amplifiers, *Electron. Lett.*, **34**, 87-88 (1998).
32. L. Gillner, Properties of optical switching networks with passive or active space switches *IEE Proc. Pt. J. Optoelectronics*, **140**, 309-315 (1993).
33. R.F. Kalman, L.G. Kazovsky and J.W. Goodman, Space division switches based on semiconductor optical amplifiers, *IEEE Photon. Technol. Lett.*, **4**, 1048-1051 (1992)
34. A. Ehrhardt, M. Eiselt, G. Grossopf, L. Kuller, R. Ludwig, W. Pieper, R. Schnabel, and H.G. Weber, Semiconductor laser amplifier as optical switching gate, *J. Lightwave Technol.*, **11**, 1287-1295 (1993).
35. J. Yu and P. Jeppesen, Improvement of cascaded semiconductor optical amplifier gates by using holding light injection, *J. Lightwave Technol.* **19**, 614-623 (2001).
36. J-H. Kim, K-R. Oh, H-S. Kim and K. Cho, All-optical switching by counterpropagating operation in cascaded semiconductor optical amplifiers, *IEEE Photon. Technol. Lett.*, **12**, 513-515 (2000).
37. G. Sherlock, J.D. Burton, P.J. Fiddyment, P.C. Sully, A.E. Kelly and M.J. Robertson, Integrated 2x2 optical switch with gain, *Electron. Lett.*, **30**, 137-138 (1994).
38. R.S. Fan and R.B. Hooker, Hybrid optical switch using passive polymer waveguides and semiconductor optical amplifiers, *J. Lightwave Technol.*, **18**, 546-554 (2000).

39. H. Ishii, M. Kohtoku, Y. Shibata, S. Oku, Y. Kadota, Y. Yoshikuni, H. Sanjoh, Y. Kondo, and K. Kishi, Zero insertion loss operations in monolithically integrated WDM channel selectors, *IEEE Photon. Technol. Lett.*, **11**, 242-244 (1999).
40. M.A. Franke and H.G. Winful, All-optical switching in an angled-grating semiconductor Bragg amplifier, *IEEE Photon. Technol. Lett.*, **11**, 815-817 (1999).
41. J. Leuthold, P.A. Besse, J. Eckner, E. Gamper, M. Dulk and H. Melchior, All-optical space switches with gain and principally ideal extinction ratios, *IEEE J. Quantum Electron.*, **34**, 622-633 (1998).
42. S. Nakamura, Y. Ueno, K. Tajima, J. Sasaki, T. Sugimoto, T. Kato, T. Shimoda, M. Itoh, H. Hatakeyama, T. Tamanuki and T. Sasaki, Demultiplexing of 168-Gb/s data pulses with a hybrid-integrated symmetric Mach-Zehnder all-optical switch, *IEEE Photon. Technol. Lett.*, **12**, 425-427 (2000).
43. R.J. Manning, A.E. Kelly, A.J. Poustie and K.J. Blow, Wavelength dependence of switching contrast ratio of semiconductor optical amplifier-based nonlinear loop mirror, *Electron. Lett.*, **34**, 916-918 (1998).
44. J-H Lee, D-A. Wang, Y-W. Kiang, H-J. Chiang, D-W. Huang and C.C. Yang, Nonlinear switching behaviours in a compact all-semiconductor optical-amplifier Sagnac interferometer device *IEEE J. Quantum Electron.*, **35**, 1469-1477 (1999).
45. R. van Roijen, J.M.M. van der Heijden, L.F. Tiemeijer, P.J.A. Thijs, T. van Dongen, J.J.M. Binsma and B.H. Verbeek, Over 15 dB gain from a monolithically integrated optical switch with an amplifier, *IEEE Photon. Technol. Lett.*, **5**, 529-531 (1993).
46. H. Soto, D. Erasme and G. Guekos, 5-Gb/s XOR optical gate based on cross-polarization modulation in semiconductor optical amplifiers, *IEEE Photon. Technol. Lett.*, **13**, 335-337 (2001).
47. K. Vahala, R. Paiella and G. Hunziker, Ultrafast WDM logic, *IEEE J.* Sel. Top. in Quantum Electron., **3**, 698-701 (1997).
48. T. Houbavlis, K. Zoiros, A. Hatziefremidis, H. Avramopoulos, L. Occhi, G. Guekos, S. Hansmann, H. Burkhard and R. Dall'Ara, 10 Gbit/s all-optical Boolean XOR with SOA fibre Sagnac gate, *Electron. Lett.*, **35**, 1650-1652 (1999).
49. T. Fjelde, D. Wolfson, A. Kloch, C. Janz, A. Coquelin, I. Guillemot, F. Gaborit, F. Poingt, B. Dagens and M. Renaud, 10 Gbit/s all-optical logic OR in monolithically integrated interferometric wavelength converter, *Electron. Lett.*, **36**, 813-815 (2000).
50. A. Sharaiha, H.W. Li, F. Marchese and J. Le Bihan, All-optical logic NOR gate using a semiconductor laser amplifier, *Electron. Lett.*, **33**, 323-325 (1997).
51. K. Uchiyama, S. Kawanishi and M. Saruwatari, 100-Gb/s multiple-channel output all-optical OTDM demultiplexing using multichannel four-wave mixing in a semiconductor optical amplifier, *IEEE Photonics Technol. Lett.*, **10**, 890-892 (1998).
52. St. Fischer, M. Dulk, E. Gamper, W. Vogt, W. Hunziker, E. Gini, H. Melchior, A. Buxens, H.N. Poulsen and A.T. Clausen, All-optical regenerative OTDM add-drop multiplexing at 40 Gb/s using monolithic InP Mach-Zehnder interferometer, *IEEE Photon. Technol. Lett.*, **12**, 335-337 (2000).
53. E. Jahn, N. Agrawal, W. Pieper, H.-J. Ehrke, D. Franke, W. Furst and C.M. Weinert, Monolithically integrated nonlinear Sagnac interferometer and its application as a 20 Gbit/s all-optical demultiplexer, *Electron. Lett.*, **32**, 782-784 (1996).
54. D. Zhou, K. Il Kang I. Glesk, and P.R. Prucnal, An analysis of signal-to-noise ratio and design parameters of a terahertz optical asymmetric demultiplexer, *J. Lightwave Technol*, **17**, 298-307 (1999).

55. G. Raybon, U. Koren, B.I. Miller, M. Chien, M.G. Young, R.J. Capik, K. Dreyer and R.M. Derosier, A wavelength-tunable semiconductor amplifier/filter for add/drop multiplexing in WDM networks, *IEEE Photon. Technol. Lett.*, **9**, 40-42 (1997).
56. M.W.K. Mak, H.K. Tsang, H.F. Liu, Wavelength-tunable 40 GHz pulse-train generation using 10 GHz gain-switched Fabry-Perot laser and semiconductor optical amplifier, *Electron. Lett.*, **36**, 1580-1581 (2000).
57. H. Shi, J. Finlay, G.A. Alphonse, J.C. Connolly and P.J. Delfyett, Multiwavelength 10-GHz picosecond pulse generation from a single-stripe semiconductor diode laser, *IEEE Photon. Technol. Lett.*, **9**, 1439-1441 (1997).
58. O. Kamatani and S. Kawanishi, Prescaled timing extraction from 400 Gb/s optical signal using a phase lock loop based on four-wave-mixing in a laser diode amplifier, *IEEE Photon. Technol. Lett.*, **8**, 1094-1096 (1996).
59. H.J. Lee, H.G. Kim, J.Y. Choi and H.K. Lee All-optical clock recovery from NRZ data with simple NRZ-to-PRZ converter based on self-phase modulation of semiconductor optical amplifier, *Electron. Lett.*, **35**, 989-990 (1999).
60. T. Yamamoto, L.K. Oxenlowe, C. Schmidt, C. Schubert, E. Hilliger, U. Feiste, J. Berger, R. Ludwig and H.G. Weber, Clock recovery from 160 Gbit/s data signals using phase-locked loop with interferometric optical switch based on semiconductor optical amplifier *Electron. Lett.*, **37**, 509-510 (2001).
61. R.J. Manning, I.D. Phillips, A.D. Ellis, A.E. Kelly, A.J. Poustie and K.J. Blow, All-optical clock division at 40 GHz using semiconductor optical amplifier based nonlinear interferometer, *Electron. Lett.*, **35** 827-829 (1999).
62. U. Feiste, R. Ludwig, E. Dietrich, S. Diez, H.J. Ehrke, D. Razic and H.G. Weber, 40 Gbit/s transmission over 434 km standard fibre using polarisation independent mid-span spectral inversion, *Electron. Lett.*, **34**, 2044-2045 (1998).
63. K. Kikuchi and K. Matsuura, Transmission of 2-ps optical pulses at 1550 nm over 40-km standard fiber using midspan optical phase conjugation in semiconductor optical amplifiers, *IEEE Photon. Technol. Lett.*, **10**, 1410-1412 (1998).
64. M.F.C. Stephens, D. Nesset, K.A. Williams, R.V. Penty, I.H. White, and M.J. Fice, Dispersion compensation at 40 Gbit/s over 100 km of standard fibre via mid-span spectral inversion in semiconductor optical amplifier with integrated pump laser, *Electron. Lett.*, **35**, 1359-1361 (1999).
65. M. Gustavsson, A. Karlsson and L. Thylen, Travelling wave semiconductor laser amplifier detectors, *J. Lightwave Technol*, **8**, 610-617 (1990).
66. M.J. Connelly and R.F. O'Dowd, Travelling wave semiconductor laser amplifier detector noise characteristics, *IEE Proc. Pt. J. Optoelectron.*, **142**, 23-28 (1995).
67. R.M. Fortenberry, A.J. Lowery and R.S. Tucker, Up to 16 dB improvement in detected voltage using two-section semiconductor optical amplifier detector, *Electron. Lett.*, **28**, 474-476 (1992).

# Index

9 780792 376576

GPSR Compliance

*The European Union's (EU) General Product Safety Regulation (GPSR) is a set of rules that requires consumer products to be safe and our obligations to ensure this.*

*If you have any concerns about our products, you can contact us on ProductSafety@springernature.com*

In case Publisher is established outside the EU, the EU authorized representative is:

Springer Nature Customer Service Center GmbH
Europaplatz 3
69115 Heidelberg, Germany

**Batch number: 10755292**

Printed by Printforce, the Netherlands